NAMING OUR ANCESTORS

NAMING OUR ANCESTORS
An Anthology of Hominid Taxonomy

W. Eric Meikle
Institute of Human Origins

Sue Taylor Parker
Sonoma State University

WAVELAND
PRESS, INC.

Prospect Heights, Illinois

For information about this book, write or call:

Waveland Press, Inc.
P.O. Box 400
Prospect Heights, Illinois 60070
(708) 634-0081

We would like to express our gratitude to the following publishers for their
kind permission to reproduce articles in this book:

Macmillan Magazines Ltd. (*Nature*); E. Schweizerbart'sche
Verlagsbuchhandlung (*Zeitschrift für Morphologie und Anthropologie*);
South African Journal of Science; *Časopis pro mineralogii a geologii*;
The Cleveland Museum of Natural History (*Kirtlandia*); Progress
Publishers; CSHSQB Publications (*Cold Spring Harbor Symposia on
Quantitative Biology*); Royal Anthropological Institute of Great Britain
and Ireland; Harcourt Brace Jovanovich Ltd. (*Journal of Human
Evolution*)

Contents

vi   Contents

Preface

Reading the original literature of the past is an important, if often neglected, part of a scientific education. Secondary literature, and especially introductory textbooks, seem inevitably to provide a homogenized view of a field and only limited versions of its history and development. Original publications can provide the authenticity and liveliness often missing in secondary accounts. However, students generally do not have easy access to much of the literature. Many significant publications are old and/or rare, and therefore not among the holdings of most libraries. Access to many publications is limited, even where they are available in some form. They may exist only on microfiche, or be housed in off-campus collections, or require borrowing through Interlibrary Loan. They may have been published in languages which students don't read.

Reprint collections can be guides to significant literature. Some gather together important contemporary papers. Others bring together older materials, now largely of historical rather than immediate research interest. In the study of nomenclature, however, publications of any age are of more than historical interest because of their continuing relevance for determining and maintaining the correct names for taxa. This anthology makes available an important set of reprinted publications, many of which are difficult for students or instructors to obtain for reasons cited above.

Our collection contains papers from that circumscribed area of the paleoanthropological literature dealing with the scientific names of fossil species. This area is, in turn, only a very small portion of the field of zoological nomenclature, the naming of animal species (and other taxa), both living and extinct. Although relatively brief, this book includes key documents from the literature of human evolution relevant to the history of hominid taxonomy and to the discovery and naming of extinct species. This volume is designed as a reference book to be consulted periodically for specific historical information.

Most of the papers reprinted here were chosen for their historical significance as the sources of formal scientific names either currently or potentially in use to designate fossil hominid species. They were not chosen for their factual content or with any criteria other than their role in nomenclature. Accordingly, readers should not assume that the editors share any particular view of the fossil record just because we have reprinted it. Many statements in these papers are now known to be incorrect, or at least in need of qualification. We have not attempted to point out errors or make corrections in most such cases. Since these papers are all reprinted from elsewhere, there are no names in this book which are new for purposes of nomenclature. It goes without saying that all taxonomic references or citations should be to the original sources, not to this book.

Several papers were translated into English for this collection. Selection 2 (E. Dubois) was translated by Berkeley Scientific Translation Service. Selections 3 (O. Schoetensack), 9 (H. Weinert), and 12 (C. Arambourg and Y. Coppens) were translated by one of us (W.E.M.). We would like to thank Pam Moody and Larissa Smith for assistance with the German translations. For help in obtaining copies of various publications we thank the Interlibrary Loan offices and Reference Librarians at Sonoma State University and the University of California, Berkeley, as well as the Center for Research Libraries, Chicago. Support during the preparation of this book was provided by our respective institutions, the Institute of Human Origins and Sonoma State University. Finally, we would like to thank those teachers and colleagues who have stimulated our interest in the history of paleoanthropology, especially F. Clark Howell.

What's in a name? That which we call a rose
By any other name would smell as sweet. . . .
—*Romeo and Juliet*, Act II, Scene 1

Introduction: Names, Binomina, and Nomenclature in Paleoanthropology

A language barrier confronts those who study any subject for the first time. Each field has its own vocabulary, its particular terminology. Mastering such special languages is part of the process of acquiring an education. This is as true of paleoanthropology as any other field. Paleoanthropology, the study of the evolutionary origins and development of humans, is literally the anthropology of the past. It combines physical anthropology and prehistoric archeology in particular and is concerned with a broad consideration of human evolution in the area where anthropology, biology, and paleontology overlap. The vocabulary of paleoanthropology includes many terms derived from biology in general, especially from anatomy, taxonomy, and evolution. One aspect of this language which immediately confronts all students of the human fossil record is the naming of species. Often long, sometimes complicated and obscure, the scientific names of various entities, living or extinct, must be recognized, distinguished, and associated with their proper referents before communication about the human fossil record can proceed. Confronting "*Australopithecus*," for example, learning to spell it or to recognize and understand its roots, should lead to further questions. Where did this name come from? When and why was it created? To what does it refer? How have ideas about its components, the organisms labeled with this name,

1

changed over time? Following up on questions such as these leads to an understanding of the history of ideas about human evolution and the human fossil record. Understanding the names which have been applied to various groups, and the evolutionary relationships among these groups which are symbolized by the organization of names into classifications, is fundamental to studying paleoanthropology.

Before we consider the system of naming, responsible for such labels as "*Australopithecus*," let us distinguish a few biological terms which are often used in overlapping fashion: **systematics, classification**, and **taxonomy**. Precise definitions vary, but the classic formulations of Simpson (1961) and Mayr and Ashlock (1991) remain useful for our purposes.

Systematics is "the scientific study of the kinds and diversity of organisms and of any and all relationships among them" (Simpson: p. 7) or, more simply, "the science of the diversity of organisms" (Mayr & Ashlock: p. 2). Systematics as thus defined is a broad science which includes classification, taxonomy, ecology, and other biological disciplines within it. "Zoological classification is the ordering of animals into groups (or sets) on the basis of their relationships, that is, of associations by contiguity, similarity, or both" (Simpson: p. 9). "Classification makes organic diversity accessible to the other biological disciplines. Without it, most of them would be unable to give meaning to their findings" (Mayr & Ashlock: p. 2). The common framework of most zoological classifications is that originally made popular in the 18th century by Carolus Linnaeus, the Swedish founder of modern classification and nomenclature. This system, with its progressively more inclusive categories of species, genus, family, order, and so forth, is called "hierarchic" because each category occupies a particular rank, with more inclusive categories above and less inclusive ones below. "Taxonomy is the theoretical study of classification, including its bases, principles, procedures, and rules" (Simpson: p. 11). Again more simply, "taxonomy is the theory and practice of classifying organisms" (Mayr & Ashlock: p. 2).

The objects studied in taxonomy are **taxa** (singular: **taxon**). "A taxon is a group of real organisms recognized as a formal unit at any level of a hierarchic classification" (Simpson: p. 19). "The term *taxon* always refers to concrete zoölogical objects . . . A *category* designates rank or level in a hierarchic classification. *It is a class whose members are all the taxa that are assigned a given rank*" (Mayr & Ashlock: pp. 20–21). For example, "*Homo*" is a taxon; "genus" is a category. "Hominidae" is a taxon; "family" is a category.

The most basic unit of interest in systematics and taxonomy

is the **species**. In this regard, it is essential to distinguish between the "species category" and the "species taxon," both of which are commonly just called "species." The **species category** is a basic rank in the Linnaean hierarchy—that below genus. **Species taxa** are the particular groups of related organisms which occupy this rank. How can species taxa be recognized, properly classified, and distinguished from each other? In other words, what is a species? We have no intention of trying to resolve this question here. The literature on this topic is vast and seemingly ever growing, with basic disagreements which preclude giving a precise, generally accepted definition. For recent reviews and commentary on the species question and its relevance to primate evolution, see Ereshefsky (1992) and Kimbel and Martin (1993). In any case, the authors of the papers reprinted here certainly held a variety of species concepts, and no single definition or brief discussion could adequately represent their opinions. In practice, a species is a taxon which occupies the lowest, but most biologically significant, level or category of the Linnaean hierarchy.

Nomenclature

In order to communicate about taxonomy, classifications, or anything else regarding organisms, it is necessary to apply consistent names to them. "Zoological nomenclature is the application of distinctive names to each of the groups recognized in any given zoological classification" (Simpson, 1961: p. 9). "Zoological nomenclature is merely a labeling of the taxa of classifications. It has no other function in taxonomy. It provides a vocabulary for writing and talking about animals, and so is absolutely essential to zoology, but it has no other zoological or scientific interest in itself. It is an arbitrary device that has become an enormously complex, strictly formal, rigidly legalistic system" (Simpson, 1961: p. 34).

Linnaeus established the use of a binominal ("two-names") system, combining generic and specific names to refer to each of his recognized species. The tenth edition of his *Systema Naturae*, published in 1758, is formally taken as the beginning of modern zoological taxonomy and nomenclature. The first species listed in this classification was *Homo sapiens*, followed by other primates and then other living animals. This position reflects a long tradition, before and after Linnaeus, of humans viewing themselves as primary in interest and importance among the world's inhabitants. It was more than a century later that the first extinct hominid species was named (see selection 1). Workers following Linnaeus in the 18th and 19th centuries were not uniform in their treatment of species recognition and naming. Through those years a large number of animals and plants were being collected, given scientific names, and published for the first time by European naturalists. Confusion increased as some taxa were named several times and workers studying different animal groups followed different naming customs. By the mid-19th century it was obvious that the zoological naming process needed some sort of organization. A variety of rules were adopted at meetings of the International Congress of Zoology, starting in 1889. In 1905 these were published as the *Règles internationales de la Nomenclature zoologique*. These were further amended over the years until they were replaced in 1961 by the current International Code of Zoological Nomenclature (referred to throughout this book as "the Code"), which is now in its third edition (International Commission on Zoological Nomenclature, 1985).

The Code is designed "to provide the maximum universality and continuity in the scientific names of animals compatible with the freedom of scientists to classify animals according to taxonomic judgments" (p. xiii). The Code governs the *names* of taxa, but not their *classification* or the *ranks* in the Linnaean hierarchy in which scientists place them. It guides in creating new names and in checking the status of older ones. For further information about the Code and taxonomic publication in general, see Mayr and Ashlock (1991). A copy of the Code itself (International Commission on Zoological Nomenclature, 1985) is essential for all serious students of the taxonomy of any animal group. The following briefly summarized definitions and concepts are central to the operation of the Code. These rules apply in general, although there are technical exceptions omitted from this discussion. See the Code for precise statements.

Nomenclature is a system of names and the rules for forming and using them. **Zoological nomenclature** is the system of scientific names for animal taxa. A **taxon** is any taxonomic unit, such as a species, a family, etc. The **rank** of a taxon is the level it occupies in the zoological hierarchy. A **zoological taxon** is a naturally occurring taxon of animals, whether it has been formally named or not. For example, gelada monkeys and their ancestors had probably been living in Ethiopia for millions of years before they were recognized as a species and scientifically named by taxonomists in the 19th century. A **nominal taxon** is one which exists for purposes of nomenclature; that is, one which has an available name (see below) and is based on a name-bearing type (see below). A nominal taxon should be objectively recognizable by all workers studying a group of animals. For example, the nominal genus *Theropithecus* was created by I. Geoffroy in 1843 to provide a generic name for living gelada monkeys. A **taxonomic taxon** is subjectively defined, formed by a particular worker by grouping together all the individuals he or she thinks belong together as part of a zoological taxon; it is referred to by the valid name determined from among all the available names of the included nominal taxa (see below). In other words, a taxonomic taxon is made up of one or more nominal taxa which have been subjectively combined by a taxonomist who considers them to constitute a natural, zoological taxon. For example, in a monograph, C. J. Jolly (1972) combined the two nominal genera *Theropithecus* (living gelada monkeys) and *Simopithecus* (a group of extinct fossil monkeys) into one taxonomic genus because he concluded that they were closely related at that level. The name of this taxonomic genus is *Theropithecus*. Every taxonomic taxon takes the name of one of its included nominal taxa, but it is not conceptually the same as that nominal taxon.

The Code regulates the names of three groups of taxa: the **family group**, the **genus group**, and the **species group**. Names of higher-ranking taxa, such as orders, classes, etc., are not governed by the Code. The most common members of the family group are the superfamily, family, subfamily, and tribe. The rank of taxa in the family group is indicated by suffixes combined with a common stem, as demonstrated in this example:

Rank	Suffix	Scientific Name	Common Name
Superfamily	-oidea	Hominoidea	hominoid
Family	-idae	Hominidae	hominid
Subfamily	-inae	Homininae	hominine
Tribe	-ini	Hominini	hominin

The members of the genus group are the genus and subgenus. The members of the species group are the species and subspecies. All scientific names except for those in the species group consist of one, capitalized word; examples include Hominidae and *Australopithecus*. Names in the species group always begin with a lower-case letter. The use of italics for genus- and species-group names is customary and recommended by the Code. (It is not mandated, however, by Code rules.) The scientific name of a species is a combination of generic and specific names to form a **binominal name** (a binomen; plural: binomina), such as *Homo sapiens*. The scientific name of a subspecies consists of a **trinominal name**: *Homo sapiens sapiens*.

Each nominal taxon has a name-bearing **type**, a particular individual specimen or lower-ranking taxon, which acts as an objective standard for the use of its name. The type is sometimes said to "carry" the name because they cannot be separated once properly established. Note that the type concept applies to *nominal* taxa, not other kinds. The type of a species group taxon is a particular specimen, sometimes called a "type specimen." For example, the type specimen of the nominal species *Australopithecus africanus* is the Taung juvenile (see selection 5), which will always retain this status no matter which other specimens may be grouped with it by taxonomists. Two important terms for kinds of type specimens are **holotype** and **lectotype**. A holotype is a specimen which is designated as the bearer of a species or subspecies name at the time it is first established, in the original publication. The Taung specimen is a holotype. If no holotype is designated in the original publication of a species, a lectotype can be selected later from among the original specimens to carry the species name. The type of a genus group taxon is a species, the so-called "type species." The type species of the nominal genus *Australopithecus* is *Australopithecus africanus* Dart, 1925. The type species of *Homo* is *Homo sapiens* Linnaeus, 1758. The type of a family group taxon is a genus. The type genus of the family Hominidae is *Homo* Linnaeus, 1758.

An **available name** is a scientific name which has been properly proposed and published following the various provisions of the Code; it is "available" for use in taxonomy, but not necessarily valid for a particular taxon. An **unavailable name** is a scientific name which does not conform to some Code provision(s) (mostly having to do with proper publication and form of the name) and therefore is not available for taxonomic use. A **valid name** is the correct scientific name of a taxon. An **invalid name** is an available name which is not valid because its use would violate some provision of the Code. According to the **Principle of Priority**, the valid name of a taxon is the oldest (first published) available name given to it. Priority is based on seniority determined by publication date. Later published names may remain available, but are invalid if a prior available name exists for a taxon. Priority can be set aside (by the International Commission on Zoological Nomenclature) if it would cause confusion or destroy stability of the literature due to discovery of a little-used or long-forgotten available name which is older than a generally accepted and recognized one.

Once a name is available at any rank within the family group, genus group, or species group, it continues to be available for use at other ranks within the same group, without having to be proposed or formally established again. Within a group, therefore, a taxon can be raised or lowered in rank without having to be renamed. For example, *Paranthropus* was first named as a genus by Robert Broom in 1938 (see selection 6). Many anthropologists consider this taxon to be a subgenus of *Australopithecus*, not a separate genus, and they are free to designate it as *Australopithecus (Paranthropus)* without having to create a new subgeneric name for it. In another example, a name originally proposed as a species, such as *Homo rhodesiensis* (see selection 4), can be used as a subspecies (*Homo sapiens rhodesiensis*) while retaining its original authorship and date of publication (in this case, Woodward, 1921).

Two concepts which are sometimes confused are those of **homonyms** and **synonyms**. For a species, homonyms are two (or more) available species names which are spelled the same, but applied to different nominal taxa (with different type specimens) and (either originally or later) combined in the same genus. A **senior homonym** is the first to have been named; a **junior homonym** is the later named. According to the **Principle of Homonymy** a junior homonym cannot be the valid name of a taxon. For example, suppose the species originally named *Australopithecus africanus* Dart, 1925 (whose type specimen is from Taung, South Africa, see selection 5) and *Meganthropus africanus* Weinert, 1950 (whose type specimen is from Garusi, Tanzania, see selection 9) are both considered to belong in the genus *Australopithecus*, but to be different species. Their names would become homonyms because they would represent two different species taxa with exactly the same scientific name, *Australopithecus africanus*. The result would be considerable confusion, which is avoided through the Principle of Homonymy. In this case, if *Meganthropus africanus* is transferred to *Australopithecus*, it cannot be called *Australopithecus africanus*, but must receive a new species name (see selections 9 and 14).

Finally, **synonyms** are two (or more) different scientific names of the same rank which are used to refer to the same taxon. A **senior synonym** is the first published; a **junior synonym** is the later published. **Objective synonyms** are those based on the same name-bearing type; that is, they represent two names for the same objective thing. **Subjective synonyms** are those based on different name-bearing types. As in the following case, the types may have been combined into the same taxon by some taxonomist, whose subjective judgment it is that they represent the same taxonomic taxon, although their nominal taxa differ. Robert Broom, for example, named the species *Paranthropus robustus* in 1938 and *Paranthropus crassidens* in 1949 (see selections 6 and 7). Most, but not all, paleoanthropologists today consider the type specimens of these two nominal taxa to belong to only one actual species. By the Principle of Priority the valid name of this species is the senior (subjective) synonym, *Paranthropus robustus*. The junior synonym, *Paranthropus crassidens*, remains available, but is not valid in this combination.

Taxonomic works sometimes include a "synonymy," a listing of all the various names which have been applied to a given specimen or taxon in the past. Campbell follows this tradition in part 2 of each entry in his "List of Named Hominid Taxa" (see selection 18), including bibliographic references to the original literature so that others can follow arguments through past publications. Synonymies can become complex when the literature is large and full of disagreements, as in paleoanthropology. For more recent examples of hominid synonymies, see Szalay and Delson (1979) and Groves (1989).

The scientific name of a species consists of its combination of genus and species names. In taxonomic works it is common, although not required, for a longer version of the name which also indicates its authorship and publication date to be used. For example, *Australopithecus africanus* Dart, 1925 makes explicit reference to Raymond Dart as the creator of this species name, in 1925. Note that there is no punctuation between the species name and the author's name, but a comma between the author and date. The same convention applies to generic names as well as specific ones. *Australopithecus* Dart, 1925 and *Homo* Linnaeus, 1758 are longer versions of two hominid generic names.

Other kinds of punctuation are also used to provide information about taxonomic names. For example, the names of a genus and of a subgenus are both single, capitalized, italicized words. They can be distinguished because a subgenus name is always enclosed in parentheses, but a genus name is not. A second example: when a species is transferred from its original genus to another, that fact is indicated by enclosing the author and date in parentheses. Robert Broom named the genus *Paranthropus* and the species *Paranthropus robustus* in 1938 (see selection 6). Many later authors have considered this species to belong not in its own, separate genus but in Dart's genus *Australopithecus*, which has priority over *Paranthropus* because it was named first. The full name citation would thus be *Australopithecus robustus* (Broom, 1938).

The two independent steps of recognizing taxa (by subjective judgment) and applying names (by objective recognition) are sequential. The Code is largely concerned with determining the correct ("valid") name to apply to taxa, once they have been delimited. Before names can be assigned, however, the more difficult, necessarily subjective and individual, process of recognizing taxa must be completed. The objective step depends on the publication and naming practices of the authors who have written about a taxon. Given a particular set of names and the Code, all taxonomists should, in theory, reach the same conclusion about the valid name of a taxon.

The steps for determining the name of a fossil species would therefore include:

1. Grouping all specimens into sets such that each set (the **hypodigm** of each species) corresponds to whatever species concept one has in mind;

2. Determining which type specimens are included among the hypodigm of each species to be recognized. (You may also think of this step as listing all the nominal species contained within each taxonomic species.);

3. Finding the oldest available name, and thus the likely valid name to use for each species, *given the species composition adopted*. (Changing the hypodigm of a species, by adding or dropping specimens, can result in a different valid name, depending on whether type specimens of nominal taxa are added or dropped.); and

4. If there is no valid name for a species (because none of the included specimens has been made a type in the past, or for other reasons), then proposing a new species name and a new type specimen chosen from the hypodigm.

Systematics and Evolution

While the principles of nomenclature are solely concerned with procedures for naming species and higher taxa in a consistent, stable, and universal manner, taxonomy is concerned with discovering the relationships among species. Once discovered, these relationships guide classification of species into genera, families, and other higher taxonomic categories. Because most modern taxonomists try to make their classifications consistent with genealogical relationships, they are concerned with the following issues: 1) how to understand the evolutionary relationships among the species they are classifying, and 2) how directly to reflect these relationships in their classifications.

Evolutionary taxonomy and cladism reflect two major approaches to these issues. Both evolutionary taxonomists and cladists agree that taxonomy should be consistent with phylogeny, the evolutionary history of a lineage. However, they differ in how strictly they apply the criterion of common ancestry, sometimes called monophyly. The two schools disagree over whether classification should reflect only branching speciation events (the cladist's strict application of the monophyly criterion), or whether it should also reflect the degree of anatomical, functional, or other divergence between species (the evolutionary taxonomist's more relaxed application of monophyly). Practitioners of these two schools also use different concepts and techniques to reconstruct phylogeny (Ridley, 1986).

These apparently small differences between the two approaches are significant because they can result in different classifications. Traditional evolutionary taxonomists, for example, place chimpanzees in the family Pongidae with gorillas and orangutans, reserving the family Hominidae exclusively for humans, human ancestors, and our closest extinct collateral relatives because of the differences between humans and the great apes. Cladists may place chimpanzees in the family Hominidae along with humans and human ancestors because humans and chimpanzees shared a common ancestor after the orangutan lineage had diverged. By the same token, differences between the two schools could influence the placement of fossil human species into one, two, three, or more genera depending on the phylogenies reconstructed and the significance assigned to differences among species.

Contrasting views about the tempo of evolution (Gould and Eldredge, 1977) also influence classification. Paleoanthropologists in one camp tend to classify hominid specimens into only a few species because they think hominid evolution occurred primarily through gradual transitions within species through time. Those in another camp tend to classify the same specimens into more species because they think hominid evolution occurred primarily through geologically brief, transforming speciation events.

Paleoanthropology

The short history of paleoanthropology, still less than a century and a half old, reflects tensions among various theories of evolutionary process, changes in taxonomic and nomenclatural procedures, and the impact of an increasing rate of discovery of new fossils and new scientific technologies. We only have space here to provide the barest outline of the history of discoveries related to the human fossil record. For more detailed histories and reviews, see Bowler (1986), Lewin (1987), Reader (1988), Spencer (1982), and Trinkaus and Shipman (1993). The first finds which, in retrospect, contributed to that fossil record were mostly accidental or coincidental. Early finds, before the middle of the 19th century, sometimes went unrecognized or unpublicized because of the lack of geological or evolutionary contexts in which to place such discoveries. Because the fields of paleontology and anthropology were in their early days and poorly developed, practically no scholars were prepared to study or interpret fossils. One hundred and fifty years ago these subjects had yet to become professions and to find secure homes in either universities or museums.

From the middle of the 19th century onward, the study of hominid fossils took place against the background of an exponential increase in sites, specimens, and other evidence. Additions to knowledge were at first slow and sporadic, but have become so common that now a year rarely passes without some truly significant new discovery or advance. For purposes of a brief overview we can divide the history of paleoanthropology into fifty year periods starting in about 1850. These three periods can be respectively characterized: first, as pioneering, largely amateur, and uncertain; second, as rapidly expanding and consolidating; third, as professional, integrated with other sciences, and increasingly definite.

The time between 1850 and 1900 was the period of the first significant discoveries of hominid fossils, the general recognition of a prehistoric archeological record, and the acceptance of the importance of evolution in the history of life. The discovery of the Neandertal specimen in 1856 (selection 1) coincided closely with the ferment of interest in evolution which attended the publication of Darwin's *On the Origin of Species* in 1859. However, significant intellectual development of evolutionary studies took place in the following three decades with almost no important contribution from the human fossil record. During this time the general geological, paleontological, and archeological foundations for studying the past were being laid in Europe. Most of the fossil finds of this period were geologically young, almost all being Upper Pleistocene or Holocene in current terminology. At the time, only relative dating techniques indicating "older" or "younger" were available to measure human antiquity. Finds were largely studied by anatomists or physicians, the most scientifically prepared of those interested in the human past. Total sample size for all fossil hominids was small. Eugène Dubois (selection 2) was exceptional in this period for his techniques and results. These were decades ahead of similar efforts and discoveries by others, as demonstrated by his travel to a promising but distant overseas locality far from Europe, deliberate and prolonged search for fossils in a variety of deposits, employment of a large number of workers, and success in finding geologically ancient hominid remains far older than others known at the time.

In the period between 1900 and 1950 the number of professional, full-time workers in paleoanthropology increased from a relative handful concentrated in Europe to dozens scattered around the world. These workers were still largely trained as anatomists or paleontologists, but some anthropologists had begun to appear among them. During this time, knowledge of the European record became well-grounded in understanding of regional sequences in geology and archeology. Significant fieldwork and important discoveries also came to be expected from Asia and Africa during these decades. Geologically older fossils were recovered and recognized as originating from as far back as the Pliocene. However, considerable uncertainty about the absolute ages of many discoveries remained. Most dating was based on correlation of associated faunal remains. Work at Zhoukoudian, China, in the 1920s-1930s became a model for the kind of major, long-term, international field project which has become common since World War II. Elsewhere, the potential importance of australopithecine discoveries in Africa was apparent by 1950, but the contexts of those sites were still unclear. By the end of this period relatively good sample sizes existed for Cro-Magnon, Neandertal, and *Homo erectus* groups.

In the years since 1950 geochronological advances (especially those based on radioactive isotopes such as the radiocarbon and potassium/argon dating techniques) have given reliable ages for numerous sites; firm and clear contexts have come to anchor knowledge of the hominid fossil record at many points. There are now hundreds of scientists specializing in (or occasionally participating in) paleoanthropology as a profession, worldwide. Many of these are anthropologists trained in anatomy, paleontology, zoology, etc. The educational and economic boom which followed World War II in the United States led to the provision of secure employment for several generations of anthropologists in university positions, while also providing funding for the kind of pure, unapplied research which paleoanthropology represents. During this time well-organized and well-funded international expeditions have become common. Such expeditions both require and promote the cooperation of specialists in many sciences, who are able to pursue detailed research on many kinds of problems. Paleoanthropology has become more integrated into modern evolutionary biology and science in general. Sample sizes for some fossil species range up to many hundreds of well-preserved specimens.

Some general sequences of events seem to have recurred often in the first century of the history of paleoanthropology.

1. A find is made or announced, often an isolated or fragmentary specimen, and often with uncertain or unknown geological context.

2. Debate commences about the meaning or interpretation of the find; unresolved questions about such matters as geological age are common, and especially critical for the next stage.

3. The specimen is rejected by some as too ape-like to be human or too recent to be somehow part of the human lineage; it may be seen as either an aberrant ape or an early, extinct sideline from the ancestral human lineage. Others accept it as possibly more closely related to humans.

4. Everyone waits for better evidence, whether more specimens, more complete specimens, discoveries in better context, or preferably all of these.

This sequence can be seen in the stories of the receptions of the first Neandertal, *H. erectus*, and australopithecine specimens from Neandertal, Trinil, and Taung, respectively. A cautious response and the general questions raised were probably more justified in each of these cases than one might think from reading only textbook versions of their history, which tend to distort the intellectual climate of the time by presenting past controversies in the light of today's conclusions.

The repeating pattern outlined above is at least partly related to the ironic coincidence that there is a generally inverse relationship between the historical order of discovery of hominid taxa and their geological age (or anatomical "primitiveness" compared to living humans). Repeated rejection of new hominid taxa when they were first announced was not simply a matter of prejudice or anti-evolution bias, although these were involved as well in some cases, especially for those workers who were reluctant to extend evolution theory fully to humans. This reluctance seemingly still exists for many people, although it is now much more rare among scientists than it was during the earlier periods of the history of paleoanthropology.

The following questions or problems have been central to paleoanthropology through most or all of its history. When and where did the lineage ancestral only to living humans originate? Is this a relatively recent or ancient lineage? How closely related are humans and various nonhuman primates? Which specific nonhuman primate is most closely related to humans? What were the important factors in human origins? What were the important factors in human evolution, development, or change, following those origins? To what extent were the causes or stimulants of human origins and evolution external vs. internal? How many now-extinct hominid lineages existed in the past? How were these related, and how did they interact with each other? What were the significant morphological and behavioral characteristics of fossil hominids? How rapidly and in what fashion did various hominid lineages evolve? Interest remains high in all these questions today.

Scholarly consideration of such issues requires information from the fossil record and from modern comparative research. Such information can only be gathered, analyzed, and disseminated through studies of various taxa. The names of these taxa are central to efficient communication about them. The content of paleoanthropological research is a part of systematics. Classifications reflect the results of systematic research. Nomenclature is a method of expressing and preserving the information content of classifications.

Criteria for Papers Selected

The papers reprinted here fall into two divisions, each arranged by original publication date. The first group consists of 15 publications (from 1864 to 1986) in which scientists recognized and named fossil hominid species. In all of these cases the species *name* is applied for the first time. In some the species and type specimen themselves are also new; in others a new name is applied to a previously known taxon or specimen. Each of these 15 articles represents the creation of a new nominal taxon; they are papers which "make available" significant names in the history of paleoanthropology. Most of these 15 papers are relatively brief and reprinted in their entirety. Three, however, (by O. Schoetensack, H. Weinert, and V. Alexeev) are so long, or concerned with so many topics irrelevant to this collection, that we have reprinted only excerpts. Most of the 15 were originally published in English, but four papers were translated for this collection: those by E. Dubois, O. Schoetensack, H. Weinert, and C. Arambourg and Y. Coppens.

Why did we choose these particular papers and not others? As Campbell demonstrates in selection 18, many dozens of names have been proposed with hominid fossils as type specimens. However, while dozens of names have been published, the majority of these are irrelevant to studies of human evolution for various reasons. Some do not follow the requirements of the International Code of Zoological Nomenclature and are unavailable. Many are clearly synonyms of other names and either can or will never be correctly used as species names, now or in the future. As explained above, names may be available without ever being generally adopted, depending on the judgment of taxonomists as to the number and composition of taxa represented in the fossil record. We are not suggesting that the nominal taxa proposed in the papers we have chosen represent all the actual fossil hominid species to be expected (by any of the criteria currently in use for recognizing fossil mammalian species). Some of them are likely to be synonyms (although opinions would differ about which ones these are). In addition, new species of fossil hominids undoubtedly await future publication; some are undiscovered, some unrecognized or at least unnamed.

Leaving aside such hypothetical or potential taxa, we have tried to include all the fossil species accepted according to a broad range of informed opinion among paleoanthropologists (such as *Homo erectus*, *Australopithecus africanus*, *Australopithecus boisei*); some species not universally recognized but with recent, serious proponents (such as *Homo neanderthalensis*, *Australopithecus aethiopicus*, *Homo ergaster*); and some nominal species with available names in the literature which may be more commonly employed in the future, depending on changes in opinion and new discoveries (such as *Homo heidelbergensis*, *Homo rhodesiensis*). Note that in the Contents and editors' notes before each selection we have referred to each nominal taxon by the name originally given it by its author(s). Many of these taxa are best known today in a different combination of genus and species names or by the name of a senior subjective synonym. We hope that referring to the full original name will help readers recognize such generic transfers and other nomenclatural changes in the original literature.

At the beginning of our editors' comments on each reprinted paper we have also indicated the type specimen for the species first published there. We have used commonly accepted specimen designations, mainly following the *Catalogue of Fossil Hominids* (Oakley et al., 1971; 1975; 1977). Early in the history of paleoanthropology, when relatively few specimens were known, many fossils were referred to by scientific name (with a distinct genus and/or species for each specimen), or only by site name or another informal term. Today anthropologists appreciate the need to refer to individual specimens by some formal numbering system and to reserve use of Latin, scientific names for reference to taxa, thus avoiding the misleading evolutionary implications of needlessly assigned generic and specific names. For a discussion of the importance of referring to objects with the correct terminology, see Simpson (selection 17).

Each of these reprinted papers must be considered in historical context. The phrasing, attitudes, assumptions, and so forth of some authors may seem archaic, naive, or objectionable today. However, such judgments would be unfair. Only a "whig interpretation of prehistory," with apologies to H. Butterfield (1951), insists on denigrating past opinions which are seen in retrospect as flawed merely because they do not correspond to our current views. Nothing is more certain in paleoanthropology than the discovery of new fossils and sites, some of which are certain to cause reassessments, whether minor or major, of scientific conclusions which are today widely accepted.

The second group of papers consists of four essays, published between 1950 and 1986. These represent many of the revisions and changes in taxonomic practice which have taken place in paleoanthropology since World War II. Obviously the choice of papers in this second group is more subjective than in the first. Each of the second group of papers (by E. Mayr, G. G. Simpson, B. G. Campbell, and I. Tattersall) treats important issues in the nomenclature and/or taxonomy of fossil hominids. Each is relevant to a different aspect of the naming of fossil species. Mayr (1950) applies systematic concepts and assumptions which were new to anthropology to the data of the fossil record and suggests drastic changes in the hominid taxonomy of the day. Simpson (1963) reviews the bases of taxonomy and points out many of the faulty practices of earlier paleoanthropologists. Campbell (1965) reviews in detail the status of names previously applied to fossils. Tattersall (1986) challenges the result of the "lumping" tradition begun by Mayr's paper as too extreme. The further significance of each essay is briefly discussed in the editors' notes preceding it.

Literature Cited*

Arambourg, C. & Coppens, Y. (1967). Sur la découverte, dans le Pléistocène inférieur de la vallée de l'Omo (Éthiopie), d'une mandibule d'Australopithécien. *Comptes Rendus Académie Sciences Paris, série D, 265*(8): 589–590.

Bowler, P. J. (1986). *Theories of Human Evolution*. Baltimore: Johns Hopkins University Press.

Butterfield, H. (1951). *The Whig Interpretation of History*. New York: Charles Scribner's Sons.

Day, M. H., Leakey, M. D., & Olson, T. R. (1980). On the status of *Australopithecus afarensis*. *Science, 207*: 1102–1103.

Day, M. H. & Molleson, T. I. (1973). The Trinil femora. In M. H. Day (Ed.), *Human Evolution* (pp. 127–154). London: Taylor & Francis.

Ereshefsky, M. (Ed.) (1992). *The Units of Evolution: Essays on the Nature of Species*. Cambridge: MIT Press.

Feibel, C. S., Brown, F. H., & McDougall, I. (1989). Stratigraphic context of fossil hominids from the Omo Group deposits: Northern Turkana

---

*Please note that works cited in the editors' commentaries preceding each reading are included in this list.

Basin, Kenya and Ethiopia. *American Journal of Physical Anthropology*, 78: 595–622.

Findlay, G. (1972). *Dr. Robert Broom, F.R.S.* Cape Town: A. A. Balkema.

Gould, S. J. & Eldredge, N. (1977). Punctuated equilibria: Tempo and mode of evolution reconsidered. *Paleobiology*, 3: 115–151.

Grine, F. E. (1985). Dental morphology and the systematic affinities of the Taung fossil hominid. In P. V. Tobias (Ed.), *Hominid Evolution: Past, Present and Future* (pp. 247–253). New York: Alan R. Liss.

Grine, F. E. (Ed.) (1988a). *Evolutionary History of the "Robust" Australopithecines.* New York: Aldine de Gruyter.

Grine, F. E. (1988b). Evolutionary history of the "robust" australopithecines: A summary and historical perspective. In F. E. Grine (Ed.), *Evolutionary History of the "Robust" Australopithecines* (pp. 509–520). New York: Aldine de Gruyter.

Groves, C. P. (1989). *A Theory of Human and Primate Evolution.* Oxford: Oxford University Press.

Hinrichsen, D. (1978). How old are our ancestors. *New Scientist*, 78: 571.

International Commission on Zoological Nomenclature (1985). *International Code of Zoological Nomenclature* (3rd ed.). Berkeley: University of California Press.

Johanson, D. C. & White, T. D. (1980). On the status of *Australopithecus afarensis. Science*, 207: 1104–1105.

Jolly, C. J. (1972). The classification and natural history of *Theropithecus (Simopithecus)* (Andrews, 1916), baboons of the African Plio-Pleistocene. *Bulletin of the British Museum (Natural History), Geology*, 22(1): 1–123.

Kimbel, W. H. & Martin, L. B. (Eds.) (1993). *Species, Species Concepts, and Primate Evolution.* New York: Plenum Press.

Leakey, M. D. & Harris, J. M. (Eds.) (1987). *Laetoli: A Pliocene Site in Northern Tanzania.* Oxford: Clarendon Press.

Lewin, R. (1987). *Bones of Contention.* New York: Simon and Schuster.

Mayr, E. & Ashlock, P. D. (1991). *Principles of Systematic Zoology* (2nd ed.). New York: McGraw-Hill.

Oakley, K. P., Campbell, B. G., & Molleson, T. I. (Eds.) (1971). *Catalogue of Fossil Hominids. Part II: Europe.* London: British Museum (Natural History).

_____. (1975). *Catalogue of Fossil Hominids. Part III: Americas, Asia, Australasia.* London: British Museum (Natural History).

_____. (1977). *Catalogue of Fossil Hominids. Part I: Africa* (2nd ed.). London: British Museum (Natural History).

Olson, T. R. (1974). Taxonomy of the Taung skull. *Nature*, 252: 85.

Protsch, R. R. R. (1981). *Die Archäologischen und Anthropologischen Ergebnisse der Kohl-Larsen-Expeditionen in Nord-Tanzania, 1933–1939. Band 3. The Palaeoanthropological Finds of the Pliocene and Pleistocene.* Tübingen: Verlag Archaeologica Venatoria.

Ravosa, M. J. (1991). Interspecific perspective on mechanical and nonmechanical models of primate circumorbital morphology. *American Journal of Physical Anthropology*, 86: 369–396.

Reader, J. (1988). *Missing Links* (2nd ed.). New York: Penguin Books.

Ridley, M. (1986). *Evolution and Classification*. London: Longman.

Senyürek, M. (1955). A note on the teeth of *Meganthropus africanus* Weinert from Tanganyika Territory. *Türk Tarih Kurumu Belleten, 19*(73): 1–55.

Simpson, G. G. (1961). *Principles of Animal Taxonomy*. New York: Columbia University Press.

Spencer, F. (Ed.) (1982). *A History of American Physical Anthropology, 1930–1980*. New York: Academic Press.

Szalay, F. S. & Delson, E. (1979). *Evolutionary History of the Primates*. New York: Academic Press.

Theunissen, B. (1989). *Eugène Dubois and the Ape-Man from Java: The History of the First "Missing Link" and Its Discoverer*. Dordrecht: Kluwer Academic Publishers.

Tobias, P. V. (1967). *Olduvai Gorge, Volume 2: The Cranium and Maxillary Dentition of* Australopithecus (Zinjanthropus) boisei. Cambridge: Cambridge University Press.

_____. (1973). Implications of the new age estimates of the early South African hominids. *Nature, 246*: 79–83.

_____. (1974). Taxonomy of the Taung skull. *Nature, 252*: 85–86.

_____. (1978). The South African australopithecines in time and hominid phylogeny, with special reference to the dating and affinities of the Taung skull. In C. J. Jolly (Ed.), *Early Hominids of Africa* (pp. 45–84). London: Duckworth.

_____. (1984). *Dart, Taung and the "Missing Link"*. Johannesburg: Institute for the Study of Man in Africa.

_____. (1991). *Olduvai Gorge, Volume 4: The Skulls, Endocasts and Teeth of* Homo habilis. Cambridge: Cambridge University Press.

Trinkaus, E. & Shipman, P. (1993). *The Neandertals*. New York: Alfred A. Knopf.

White, T. D. (1981). Primitive hominid canine from Tanzania. *Science, 213*: 348–349.

Wood, B. (1991). *Koobi Fora Research Project, Volume 4: Hominid Cranial Remains*. Oxford: Clarendon Press.

Part I

Diagnoses of Hominid Species

1

Homo neanderthalensis King, 1864
Type specimen: Neandertal 1

We begin with the first naming, in print, of a fossil "human" species, *Homo neanderthalensis*, by William King in 1864. The Neandertal (or "Neanderthal"; the "h" is silent and both spellings continue to be used by anthropologists) specimen was found in Germany in 1856 (not 1857, as King states). It was the first fossil hominid specimen to be widely recognized, when discovered, as morphologically non-modern compared to living *Homo sapiens*. King's paper was not the first to consider this partial skeleton; he cites previous publications by others. However, it was the first to propose a new species name for the specimen, even if the name is mentioned in the final footnote as if merely an afterthought.

Controversy about the significance of the Neandertal specimen continued for years because essentially no fossil record of early humans was known with which to compare this find and because the evidence of its geological age was ambiguous. Given only living primates as a source of anatomical comparison and almost no geological evidence of its antiquity, a wide range of opinions about the specimen was possible. These debates were finally resolved decades later with the recovery in Europe of additional, similar, fossil specimens of what came generally to be called "Neandertals."

Most of King's paper consists of anatomical description and comparison of the skull. He points out many features in which it differs from living humans and resembles apes. Today, given the accumulation of scientific knowledge over more than a century, Neandertals seem not at all ape-like. However, given the contemporary state of knowledge and the choice only of living humans or apes for comparison, it is not surprising that "ape-like" features of the first Neandertal were emphasized. The browridges were especially significant in this regard. Speculations about browridges and interpretations of their function(s) have been seemingly endless over the decades and remain a topic of current research (Ravosa, 1991).

The anatomical distinctions between Neandertals and recent humans pointed out by King were confirmed and supplemented over the next fifty years by discoveries of more fossils elsewhere in Europe. As a result, Neandertals came to be accepted as a distinct, anatomically non-modern, hominid group, the first recognized as separate from, although related in some fashion to, recent *Homo sapiens*. Whether considered in some way ancestral to modern humans, or only an extinct branch away from our lineage, Neandertals remain critical to understanding human evolution over the last few hundreds of thousands of years. Virtually all later workers have followed King in recognizing the Neandertals as a distinct taxon. The name *neanderthalensis* has been universally retained as a species-group name within the genus *Homo*, whether as a species (*Homo neanderthalensis*) or a subspecies (*Homo sapiens neanderthalensis*). The Code cannot determine whether Neandertals are a separate species or merely a temporal and geographic subspecies of *Homo sapiens*; however, it does allow us to determine the correct name formulation in either case. As originally published, the combination was *Homo neanderthalensis* King, 1864.

The Reputed Fossil Man of the Neanderthal

William King

As it is my intention to confine myself to the consideration of the Neanderthal fossil with reference to its place in Nature, I must necessarily be brief in my remarks on the circumstances under which it occurred, and on its geological age.

The fossil was found in 1857, embedded in mud in a cave or fissure intersecting the southern rocky side of the ravine or deep narrow valley, called the Neanderthal, situated near Hochdal between Düsseldorf and Elberfeld. A small stream or rivulet, known as the Düssel, flows along a narrow channel about sixty feet below the lowest part of the fissure, and on one side of the valley.

It has long been known that human bones, belonging to an extinct

From William King, "The Reputed Fossil Man of the Neanderthal," *Quarterly Journal of Science*, pp. 88–97, Vol. 1 (1864).

race, and occurring in stalagmite along with the remains of the mammoth and other fossil animals, have been found in the limestone fissures or caverns of the lofty precipices which overhang the river Meuse, in Belgium, about seventy English miles south-west of the Neanderthal.

Lyell's late work, 'The Antiquity of Man,' contains a very lucid description of the Meuse caverns, and of the one under consideration. In both cases it is evident that we have examples of ancient swallow-holes, into which have been washed bones, mud, and gravel, when their openings existed in the bed of large and powerful rivers. It was doubtless by the incessant abrading action of such ancient streams, continued for countless ages, that the Neanderthal, and much of the broad valley of the Meuse, became scooped out.

Few Geologists will dispute that the Meuse caverns are of the same age as the flint-implement gravels of the Somme, and that both belong to the latest division of the glacial or (as I have lately termed it) Clydian period.[1] If we accept the physical conditions of the Meuse caverns as demonstrative of their having been filled up in that remote age, we cannot but recognize in the corresponding conditions of the Neanderthal fissure evidences which claim for it an equally high antiquity, notwithstanding certain differences seemingly supporting the opposite conclusion.

The want of stalagmite and the *doubtful* absence of remains of extinct animals in the Neanderthal fissure may be readily explained; and as to the physical differences, the Düssel is certainly not to be compared with the Meuse for size and abrading power, but it must be admitted that a mere rivulet may take quite as much time to scoop out a "ravine" as a river to excavate a considerable portion of a broad valley.

Having finished my preliminary remarks, I shall next proceed to notice the fossil itself.

According to Dr. Fuhlrott, of Elberfeld, the skeleton was found by some workmen while quarrying the rock where the cave occurs; but, knowing nothing of the importance of the discovery, and being very careless about it, they secured chiefly only the larger bones. Fortunately these fell into the hands of Fuhlrott, and they were shortly afterwards described by Professor Schaaffhausen, of Bonn. The principal parts of the skeleton which have been preserved are the cranium; both thigh bones, perfect; a perfect right humerus; a perfect radius; the upper third of a right ulna corresponding to the humerus and radius; a left humerus, of which the upper third is wanting; a left ulna; a left ilium, almost perfect; a fragment of the right scapula; the anterior extremities of a rib of the right side; the same part of a rib of the left side; the hinder part of a rib of the right side; and two short hinder portions, and one middle portion of some other ribs.

The skeleton, or rather, as much as is preserved of it, is characterized by unusual thickness, and a great development of all the elevations and depressions for the attachment of the muscles. The ribs, which have a singularly rounded shape, and an abrupt curvature, more closely resemble the corresponding bones of a carnivorous animal, than those of man.[2]

Although a difficulty may be felt in resting a satisfactory argument upon merely the great size of its osseous framework, and the peculiar form of its ribs, it cannot but be admitted that these characters afforded some grounds for the belief at first entertained, that the Neanderthal fossil had not belonged to a human being. Whether a more close examination of other parts of the fossil will confirm this hypothesis, it is the object of the present paper to determine.

The skull is deficient in its basal and facial portions, but retains all the parts lying above a line connecting the *glabella*—or space between the eye-brows—and the *centre* of the posterior part of the skull immediately above the hollow of the neck, to which the name occipital or posterior tubercle is given.[3] Fortunately the parts alluded to, which are of uncommon thickness, enable one to determine some highly important points in craniology.

The *frontal*—or bone of the forehead[4]—possesses the upper border and roof-plate of the eye-sockets, the inter-orbital space, the orifices of the frontal sinuses, and both outer orbital processes: the upper part of the alisphenoid belonging to the right side appears also to be present. The *occipital*—or posterior bone—retains, in addition to the tubercle, the superior transverse ridges. The *parietals*—or upper side-bones—possess the impression of the temporal squamosal. The *temporals*—or lower side-bones—are broken off, though it would appear from Huxley's figure[5] that the mammillary portion of the left one is still preserved. The *lambdoidal suture*—or joining of the parietals and the occipital—including the *additamentum*, is well-marked; the *sagittal suture*—or joining of the parietals in the medio-longitudinal line of the skull—is obscure; while the *coronal suture*—or joining of the frontal and parietals in front of, and at right angles to the last-named *suture*—is but faintly marked at the crown and obliterated at the sides. The bounding line of the temporal muscles (situated on each side of the skull in front of, and above the ear) is tolerably well defined.

In general terms, the Neanderthal skull is of an elongated oval form, with a basal outline bearing much resemblance to that of the Negro cranium represented by Martin.[6] It is of large size, being about an inch longer than ordinary British skulls; in width, however, it does not much exceed them. The forehead, uncommonly low and retreating, terminates in front by enormously projecting brow or superciliary ridges, which, besides being very thick, slightly rounded on their anterior aspect, and rather strongly arched above the eye-sockets, extend uninterruptedly across from one side to the other. The outer orbital processes—or horns of the brow-ridges—are also unduly developed; being thick and projecting. On the whole, there is a remarkable absence of those contours and proportions which prevail in the forehead of our species; and few can refuse to admit that the deficiency more closely approximates the Neanderthal fossil to the anthropoid apes than to *Homo sapiens*.

The greatest width of the skull is towards its posterior part, and on a level not much higher than the mammillary region—a character which is essentially pithecoid or simial. In human skulls, the greatest width is considerably higher—usually on a line connecting the centres of ossification of the parietals:[7] on the contrary, the Neanderthal cranium, like that of the Chimpanzee, is without any particular prominency where those centres may be assumed to be situated.

In addition to possessing a low retreating forehead, the fossil skull is remarkably flattened at the vertex, which, according to Huxley, rises about 3.4 inches only above what is called the glabello-occipital plane:[8] in Man, the corresponding part is generally about an inch higher. From the vertex there is a slightly curving fall both towards the front and the back, ending in the former direction at the origin of the brow-ridges, and in the latter, at the occipital tubercle. The curving is more rounded and regular on the anterior half—particularly at the upper portion of the brow, which, in consequence, is somewhat prominent—than on the posterior half: on the latter, there is a slight depression just above the apex of the lambdoidal suture. The posterior fall of the Neanderthal skull, as a peculiarity, was first pointed out by Huxley, who remarks that "the occipital region slopes obliquely upward and forward, so that the lambdoidal suture is situated well upon the upper surface of the cranium:" in other words, when the glabello-occipital plane is made horizontal, the apex of the lambdoidal suture is decidedly in front of the posterior tubercle. In ordinary skulls, it is well known, the backward slope terminates near the apex of the lambdoidal suture, below which the occipital bone stands more or less vertical to the glabello-occipital plane. The Neanderthal cranium, in its posterior features, is approached by some savage races; also occasionally by a few inhabitants of the British Isles. Moreover, judging from the few data at our command, the approximation apparently characterized the ancient "Borreby people," and the extinct race of the Meuse, supposing the latter to be represented by a nearly perfect skull which Schmerling obtained from the Engis cave near Liège;[9] but in no human tribe extinct, or existing, do we find both the vertex and the occiput so depressed and ape-like. Well might Huxley have felt a "difficulty in believing that a human brain could have its posterior lobes so flattened and diminished as must have been the case in the Neanderthal man."

Much of the hinder half of the skull partakes of the slight roundness just noticed; but anterior to its greatest width, in the areas which were embraced by the temporal muscles, the sides are perpendicular, and their "fore and aft" outline is straight and remarkably long.

In these general characters, the Neanderthal skull is at once observed to be singularly different from all others which admittedly belong to the human species; and they undoubtedly invest it with a close resemblance to that of the young Chimpanzee, represented by Busk in his translation of Schaaffhausen's memoir.[10]

Another differential feature characterizes the fossil in question. In human skulls, even those belonging to the most degraded races, if the forehead be intersected at right angles to the glabello-occipital plane, on a line connecting the two outer orbital processes at their infero-anterior point, the intersection will cut off the frontal bone in its entire width, and to a considerable extent rising towards the coronal suture;[11] whereas in the Neanderthal skull, the same intersection will cut off only the inferior and little more than the median portion of the frontal.[12] This is quite a simial characteristic, and rarely, if ever, occurs in man.[13]

The last peculiarity is concomitant with another equally striking. Viewing the Neanderthal forehead with reference to the situation of that portion of the brain which it enclosed, we may plainly perceive that the frontal lobes of the cerebrum have been situated *behind* the outer orbital processes. As far as I have ascertained, we cannot say this of man; for, apparently, in all existing races, whose skull has not been modified by artificial pressure, the corresponding parts of the brain actually extend in *front* of the orbital processes.[14]

Notwithstanding the strong simial tendencies displayed by its general features, most of the writers who have described this skull do not appear to think otherwise than that it belonged to an individual of our species. There seems to be no doubt, whatever, on the part of the Honorary Secretary of the Anthropological Society, Mr. Carter Blake, that the Neanderthal fossil is specifically *identical* with Man. He considers it to be the remains of some poor idiot or hermit, who died in the cave where the bones were found.[15] His reasons, however, are obviously unsatisfactory. "In reply to the suggestion," observes Huxley, "that the skull is that of an idiot, it may be urged that the *onus probandi* lies with those who adopt the hypothesis. Idiotcy is compatible with very various forms and capacities of the cranium, but I know of none which present the least resemblance to the Neanderthal skull.''[16] Blake admits that its frontal peculiarities give the cranium an ''apparent ape-like character;'' but if such peculiarities be the result of mal-development producing idiotcy, one would be equally justified in believing that the form of the skull of the gorilla, or chimpanzee, is also produced by disease of the brain. Schaaffhausen, seemingly, would have no hesitation in repudiating the idea that the frontal specialities of the fossil are the result of individual pathological deformity.[17]

In case it should be suggested that this remarkable cranium has received its form from artificial pressure, I may observe that no one who has described it seems to entertain such an opinion; indeed its symmetry, also noticed by Schaaffhausen, is quite opposed to the supposition that the skull has undergone any process of artificial modification.

Huxley, while admitting that it is the most ape-like and most brutal of all human skulls yet discovered, states that it is ''closely approached'' by some Australian forms, and ''even more closely affined to the skulls of certain ancient people, who inhabited Denmark during the Stone period.''[18] I have no intention to deny that there are general features of resemblance between the Australian, Neanderthal, and ancient Danish crania; but it appears to me, judging from the figures (31 and 32) in the deeply philosophical work, 'Man's Place in Nature,' that a closer resemblance is assumed than really exists. No one would have any hesitation in admitting that the Borreby skull, represented under one of the figures cited, is strictly human,—nay, from what I have seen myself, I have no hesitation in saying that precisely the same cranial conformation is often repeated in the present day; but it has yet to be shown that any skulls hitherto found are more than *approximately* similar to the one under consideration.

The proposition at present contended for is apparently invalidated by the fact that, among certain species of animals—notably those under domestication—skulls very dissimilar from each other may be found. It is, therefore, to be apprehended that, however clearly the Neanderthal fossil may be shown to be inadmissible into the human species, an attempt will be made to set aside the consequent conclusion by an appeal to the fact alluded to. But this I contend is not a case in point, as will be evident after a moment's reflection on the various breeds of the Dog—the best known of our domesticated species. These breeds, so remarkably differentiated by cranial peculiarities, are *artificial*, whereas the varieties of mankind are *natural*. The dissimilar skulls met with in the former are merely striking illustrations of organic or structural modifiability, produced by what Darwin calls Natural Selection, but nothing more.

Again, some weight seems to be due to the consideration that the human species (in which I include all the existing races of man) is characterized by a great variety of skulls. We have abundant examples affording characters which closely link together the most dissimilar forms, so that it is impossible to draw a line of demarcation between the extremes of dolichocephaly and brachycephaly,[19] or between the lofty forehead of Indo-Europeans and the depressed one of the Australian. Nay, the most degraded race we are acquainted with—the Mincopies of the Andaman Islands—may be strictly regarded as closely affined by cranial conformation to the highest intellectual races. It might, therefore, be urged that the Neanderthal skull is simply an aberrant form, but which is, nevertheless, inseparably linked on to the Indo-European type. If sufficient has not yet been adduced to dispel this idea, the following additional evidences, referring to the particular parts of the bones composing the fossil cranium, will, it is thought, be deemed fully adequate for the purpose.

Commencing with the *Frontal.*—Fuhlrott and Huxley have satisfactorily shown that this bone is furnished with large frontal sinuses; and apparently they regard these as the cause of the excessive prominency of the superciliary ridges. It may be reasonably doubted, however, that this is the case. Frontal sinuses, it is well known, do not always coexist with prominent brow ridges, as, for example, in the Australian and the Chimpanzee: on the other hand, the former may exist without being associated with any more than an ordinary development of the latter. I have seen frontal sinuses extending to nearly the origin of the outer orbital processes, and almost large enough, even at their termination, to admit the small finger to be inserted into them, yet the brow-ridges were not particularly prominent. But whether the Neanderthal sinuses extend the whole length of the brow-ridges, or they are simply confined to the region of the *glabella*, their large size, in either case, is unusual in man, and they more strongly approach to, or resemble, as the case may be, those of the Gorilla.

As to the excessive prominency of the brow-ridges,—instead of regarding this feature as having been produced by the frontal sinuses,—there is more probability that, like the other extraordinary "elevations and depressions" of the skeleton, pointed out by Schaaffhausen, it is another speciality consequent on the greatly developed muscular system, which, from what has already been stated, evidently characterized the so-called Neanderthal man.

The orbital cavities appear to have had a circular rim, as in certain apes, there being no angle in that part joining the glabella. This is a feature unknown in any of the human races: in them the orbits are always subquadrate.[20]

The roof of the orbital cavities is altogether less concave, particularly on the outer side, than in Man; and, although the inner extremity of the plate forming the roof is broken off, sufficient remains to show that the cavities contracted sooner than usual. The cavities also *appear* to have been uncommonly divergent: if this were actually the case, its significance would point towards one of the specialities of the Gorilla.

Temporals.—As already stated, only the impression of the upper *squamosal* is seen on the parietals; but it suffices to show, as pointed out by Huxley, that this part had a comparatively low arcuation: the highest point of the arch reaches little more than half the height it attains in ordinary human skulls. Besides occurring among apes, an equally low arcuated squamosal distinguishes the human fœtus; and in some savage races—Australians and Africans—the same part is also depressed, but not so much as in the fossil. The Engis and Borreby skulls are strictly normal in this particular.[21]

Occipital.—The upper portion of this bone is quite semicircular in outline, its sutural (lambdoidal) border running with an even crescentic curve from one transverse ridge to the other:[22] generally in human skulls, including the Engis one, the outline approaches more or less to an isosceles triangle.[23] The width of the occipital at the transverse ridges is much less than is common to Man; and the disparity is the more striking in consequence of the widest portion of the fossil occupying an unusually backward position.

Taking into consideration the forward and upward curving of the upper portion of the occipital bone as previously noticed, its semicircular outline, and smallness of width, we have in these characters, taken together, a totality as yet unobserved in any human skull belonging to either extinct, or existing races; while it exists as a conspicuous feature in the skull of the Chimpanzee.

Parietals.—In Man the upper border of these bones is longer than the inferior one; but it is quite the reverse in the Neanderthal skull. The difference, amounting to nearly an inch, will be readily seen by referring to figures 1 and 2, in plate II.; the former representing the right parietal of a British human skull, and the latter the corresponding bone of the fossil. These figures also show that the Neanderthal parietals are strongly distinguished by their shape, and the form of their margins: in shape they are five-sided, and not subquadrate, like those of the British skull;[24] while their anterior and posterior margins have each exactly the reverse of the form characteristic of Man.

The *additamentum*, which undoubtedly gives the parietals their five-sided shape, is on a level with the superior transverse ridge, and much longer than usual. This peculiarity is common to the human fœtus: I have, likewise, observed an approach to it in a "Caffre" skull belonging to the Dublin University Museum, in which, also, the upper and lower borders of the parietals are about equal in length. But still the abnormality of the latter case is not at all so extreme as the condition observed in the fossil. These particular features also are characteristically simial; for in extending our survey to the Chimpanzee, and some other so-called Quadrumanes, their parietals are seen to present a great similarity to those of the Neanderthal skull.[25]

I have now, as it appears to me, satisfactorily shown that not only in its general, but equally so in its particular characters, has the fossil under consideration the closest affinity to the apes. Only a few points of proximate resemblance have been made out between it and the human skull; and these are strictly peculiar to the latter in the *fœtal state*. The cranium of the human fœtus, however, possesses the lofty dome, the forward position of the frontal respectively to the outer orbital processes, the greatest width at the parietal centres of ossification, and the vertical occipital, which are so conspicuous in the adult, but which are remarkably non-characteristic of the Neanderthal skull. Besides, so closely does the fossil cranium resemble that of the Chimpanzee, as to lead one to doubt the propriety of *generically* placing it with Man. To advocate this view, however, in the absence of the facial and basal bones, would be clearly overstepping the limits of inductive reasoning.

Moreover, there are considerations of another kind which powerfully tend to induce the belief that a wider gap than a mere generic one separates the human species from the Neanderthal fossil.

The distinctive faculties of Man are visibly expressed in his elevated cranial dome—a feature which, though much debased in certain savage races, essentially characterizes the human species. But, considering that the Neanderthal skull is eminently simial, both in its general and particular characters, I feel myself constrained to believe that the thoughts and desires which once dwelt within it never soared beyond those of the brute. The Andamaner, it is indisputable, possesses but the dimmest conceptions of the existence of the Creator of the Universe: his ideas on this subject, and on his own moral obligations, place him very little above animals of marked sagacity;[26] nevertheless, viewed in connection with the strictly human conformation of his cranium, they are such as to specifically identify him with *Homo sapiens*. Psychical endowments of a lower grade than those characterizing the Andamaner cannot be conceived to exist: they stand next to brute benightedness.

Applying the above argument to the Neanderthal skull, and considering that it presents only an approximate resemblance to the cranium of man, that it more closely conforms to the brain-case of the Chimpanzee, and, moreover, assuming, as we must, that the simial faculties are unimprovable—incapable of moral and theositic conceptions—there seems no reason to believe otherwise than that similar darkness characterized the being to which the fossil belonged.[27]

Endnotes

[1] See 'Synoptical Table of the Aqueous Rock-Systems,' 5th edition.

[2] See Busk's translation of Schaaffhausen's paper in the 'Natural History Review,' 1861, pp. 158–162.

[3] The line A A, in Figure 1, Plate I., passes from the *glabella* to the occipital tubercle.

[4] The explanation of the individual parts of the skull is prefixed to Plates I. and II.

[5] See 'Man's Place in Nature,' Fig. 25 A, facing page 138.

[6] 'Natural History of Man and Monkeys,' fig. 182, p. 120.

[7] Plate II. Figure 5, *b*.

[8] See Plate I. Figure 1, A A.

[9] This is the only speciality in which the Engis and Neanderthal skulls agree.

[10] See 'Natural History Review,' 1861, Plate IV. fig. 6.

[11] See Plate II. Figure 5, B B.

[12] See Plate I. Figure 1, B B.

[13] I have examined and made myself acquainted with skulls belonging to the principal races or varieties of man, in all of which the forward position of the forehead, relatively to the outer orbital processes, is the general rule. The Engis skull exhibits it, and the same appears to be the case with the Borreby one, judging from the figure in Lyell's 'Geological Antiquity of Man,' p. 86. It may be doubted that the Plymouth skull, represented by Busk ('Nat. Hist. Rev.' 1861, Pl. V. fig. 6), is an exception. I possess a very remarkable skull, probably about 500 years or more old, taken last summer out of the beautiful ruins of Corcomroo Abbey, situated among the Burren mountains, in county Clare, which offers a close approximation to the fossil in the depressed form of the forehead: indeed, although not altogether so abnormal in this respect as the Neanderthal skull, it has in appearance a better development, in consequence of the median part of its frontal being a little more rounded. There is no reason to believe that it belonged to an idiot, as it happens that most of the skulls lying about the ruins have a low frontal region. It is singular that the inhabitants of Burren a few hundred years ago should have been characterized by a remarkably depressed forehead, while those now living have a well-developed cranial physiognomy.

[14] The Corcomroo skull, noticed in the previous footnote, although closely approximated to the Neanderthal one in its low forehead, and *this alone*, is strictly human in the forward extension of the frontal lobes of the brain relatively to the outer orbital processes.

[15] See 'Geologist,' vol. V. p. 207.

[16] See Lyell's 'Geological Antiquity of Man,' p. 85.

[17] The writer of an article on Lyell's 'Geological Antiquity of Man,' in the last number of the 'Quarterly Review,' summarily disposes of the Neanderthal skull with the

gratuitous assertion, that it is quite removed from the pithecoid type, and possibly belonged to an idiot.

[18] 'Man's Place in Nature,' p. 157.

[19] Professor Retzius distinguished *long* skulls, and *short* or round skulls, respectively by the names *dolichocephalic* and *brachycephalic*.

[20] In some apes the rim of the orbits is of the human form.

[21] Under this head may be noticed a part which appears to have been overlooked in the fossil. On an excellent cast, supplied by Mr. Gregory, of Goldensquare, London, there occurs on the right side and in front of the squamosal impression a raised flattened plate, which looks like the upper portion of the alisphenoid (see Plate 1. Figure 1, *b*): the forward situation of this plate prevents it being taken for the anterior part of the temporal; besides, its posterior side exhibits what appears to be the impression of the squamosal. The anterior margin of the supposed alisphenoid is about an inch behind the outer orbital process. Dr. Knox long ago pointed out in a Tasmanian skull a square-shaped bone, nearly an inch in extent, interposed between the alisphenoid and the parietal. I perceive that this abnormality in a Tasmanian skull is represented in fig. 225 of the beautiful edition, just published by Renshaw, of Dr. Knox's translation of Milne-Edwards' 'Manuel de Zoologie.' I have also seen the same bone, but only on the left side, of an "Australian" skull belonging to the Dublin University Museum. Perhaps this interposed bone corresponds, in nature as well as situation, to the flattened plate observable in the Neanderthal fossil.

[22] Plate II. Figure 4.

[23] Plate II. Figure 3.

[24] The outlines were taken by pressing a sheet of paper on the parietals; and, when in this position, marking their margins by following the bounding sutures: next, by cutting the paper according to the lines given by the sutures, and allowing it to retain its acquired convexity: the outlines were then marked off on another sheet of paper. Possibly the antero-inferior angle of the Neanderthal parietal, as given in the figure, is not strictly correct, owing to the coronal suture being obliterated in that part, but I venture to state that it is approximately true.

[25] On the cast, an incised line runs from the lambdoidal suture (where the *additamentum* joins it) towards the posterior tubercle. Is this the suture which occurs near and parallel to the transverse ridges in foetal skulls, and occasionally in that of adults? In the skull of the "Caffre," noticed in the text, this suture, which is only seen on the right side, is situated above the ridge; but in the fossil, it is below this part.

[26] It has often been stated that neither the Andamaners, nor the Australians have any idea of the existence of God: there are circumstances, however, recorded of these races which prevent my accepting the statement as an absolute truth.

[27] A paper advocating the views contained in this article was read at the last meeting of the British Association, held in Newcastle-on-Tyne. In that paper I called the fossil by the name of *Homo Neanderthalensis*; but I now feel strongly inclined to believe that it is not only specifically but generically distinct from Man.

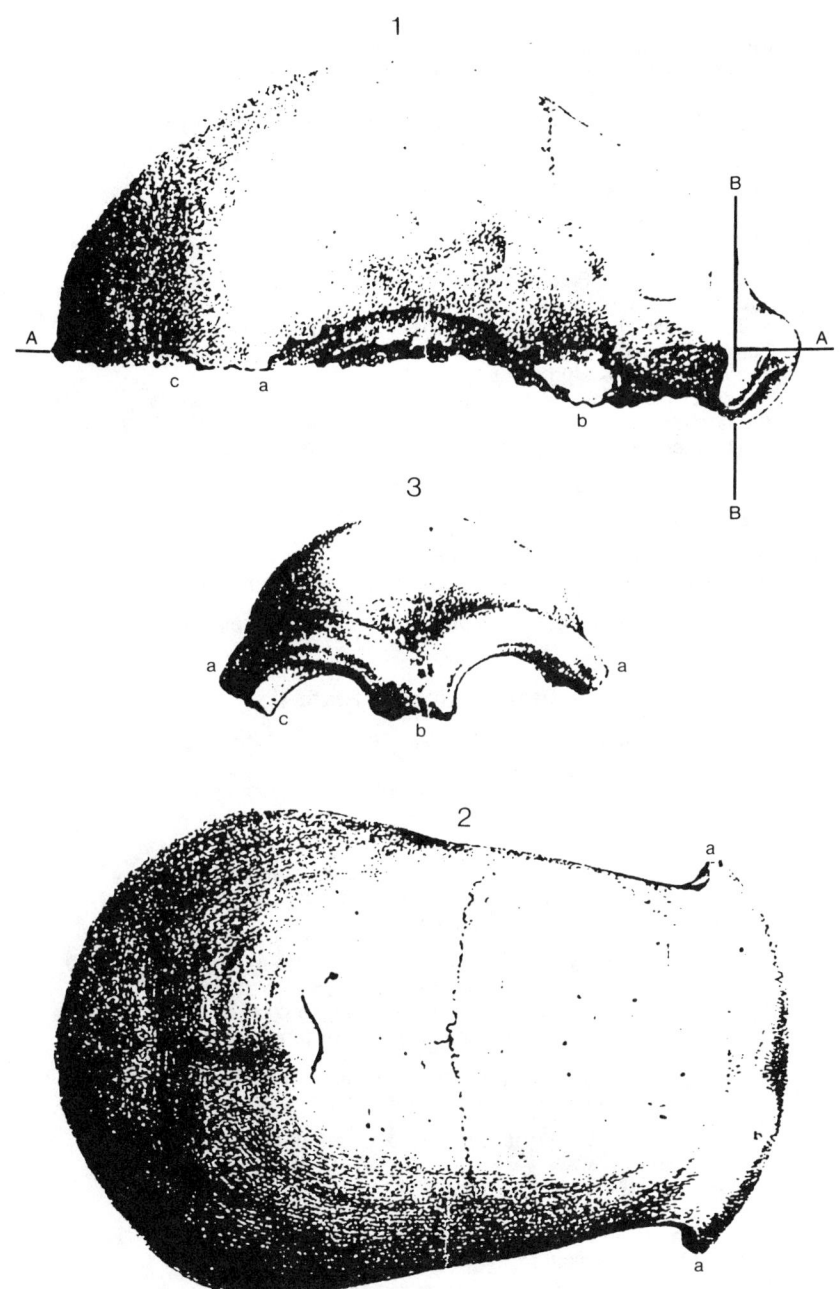

Reputed Fossil Man of Neanderthal, Plate I

Explanation of Plate I.

Figure 1.—Right Side of Neanderthal Skull.

A A. Glabello-occipital plane.

B B. Line intersecting the forehead at right angles to the last plane
through both outer orbital processes.

(These lines are interrupted so as not to obscure any parts
of the skull.)

a to a'. Border of squamosal impression.

(Letter 'a' is just below the widest part of the skull.)

b. ? Alisphenoid.

c. Portion of additamentum.

Figure 2.—Top of Neanderthal Skull.

a, a. Outer orbital processes.

The transverse line on the middle of skull represents the
coronal suture. (This and the corresponding line in Fig. 1 are
copied from Busk's figures.)

The semicircular line at the posterior part of skull represents
the lambdoidal suture.

The medio-longitudinal line represents the sagittal suture.

Figure 3.—Front of Neanderthal Skull.

a, a. Outer orbital processes or horns of the brow ridges.

b. Inter-orbital space.

c. Portion of roof-plate of right orbital cavity.

(Only the anterior half of the frontal bone is represented.)

⁎ The figures in this plate are taken from a plaster cast.

———

Explanation of Plate II.

Figure 1.—Right Parietal of a Human (Irish) Skull.
a. Coronal edge.
b. Lambdoidal edge.
c. Sagittal edge.
d. Squamosal edge.

Figure 2.—Right Parietal of Neanderthal Skull.
a, b, c, d. Same as in last Figure.
e. Additamental edge.

Figure 3.—Occipital of a Human (Irish) Skull.
a a. Lambdoidal edge.
b, b. Transverse ridges.
c. Occipital or posterior tubercle.

Figure 4.—Occipital of Neanderthal Skull.
Letters same as in last Figure.

Figure 5.—Right Side-view of Dome of Human Skull.

A A. Glabello-occipital plane.

B B. Glabello-occipital intersecting plane.

a. Frontal.

b. Parietal. (The letter is on the centre of ossification and
widest part of the skull.)

c. Occipital.

d. Temporal.

e. Alisphenoid.

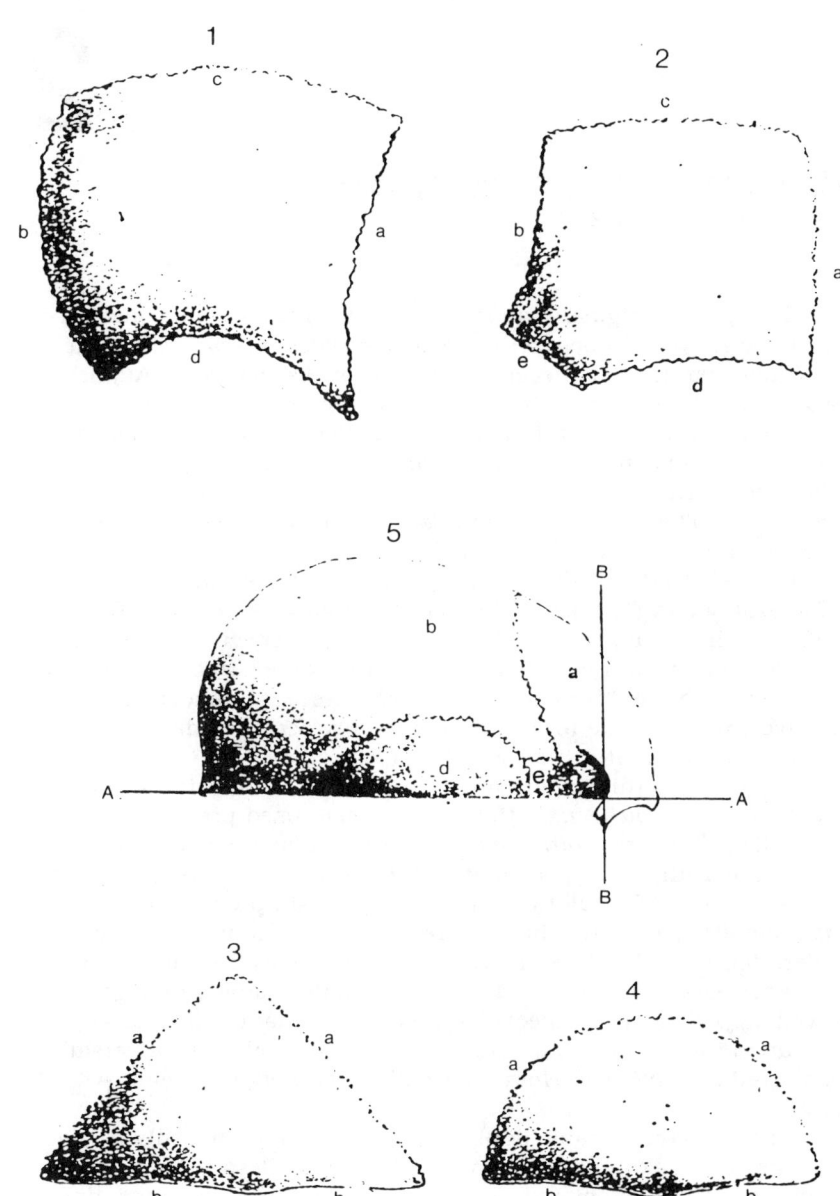

Reputed Fossil Man of Neanderthal, Plate II

2

Anthropopithecus erectus Dubois, 1892
Type specimen: Trinil 2

This article originally appeared as one of the regular quarterly
reports by Eugène Dubois on his paleontological research in what is
now Indonesia. Here he describes the femur discovered in August,
1892, at Trinil, Java, which he believed came from the same
individual as the skullcap found in 1891. For a recently published
history of Dubois and the Trinil discoveries, see Theunissen (1989).
This paper is Dubois' report for the third quarter of 1892, which,
according to Theunissen, was published in 1893. If this is correct,
the species name "*erectus* Dubois, 1892," as generally cited
(Campbell—selection 18 in this book; Groves, 1989; Oakley et al.,
1975; Szalay and Delson, 1979) should actually be "*erectus* Dubois,
1893." Such a change would have no practical effect elsewhere in
hominid nomenclature; it is simply a matter of accuracy in citation.
Because we were not able to establish the correct publication date
ourselves, we continue to use the commonly accepted date of 1892
with the caution that it may well be incorrect.

Note that in this paper Dubois still refers his species to the
genus "*Anthropopithecus*," the name he had used previously for
the skullcap. *Anthropopithecus* was a generic name commonly
used in the 19th century for living chimpanzees. Dubois did not at
first assign the 1891 calotte to a species, but simply referred to it as
Anthropopithecus. After the discovery of the anatomically very
modern femur in 1892 he created the new species first named in
this paper. Only later, in 1894, did he formally name a new genus
as well, applying Ernst Haeckel's hypothetical name *Pithecanthropus* to the Trinil finds. Today this species is all but universally
considered a member of *Homo*, formally *Homo erectus* (Dubois,
1892).

There is considerable doubt that the Trinil femur truly derives
from the same taxon as the calotte (Day and Molleson, 1973). Other
femora, certainly associated with *Homo erectus* at other sites, differ
consistently in a variety of features from the Trinil bone. However,
the possible separation of these two specimens into different taxa
has no effect on names in use because the type specimen for the

name *erectus* Dubois is the calotte, not the femur. Because Dubois indicates no type specimen in this paper a lectotype could later be designated. The calotte is viewed as the lectotype by Campbell (see selection 18) on account of its description by Dubois as "Pithecanthropus I" in 1894.

Notice again, as with King and the Neandertal find (see selection 1), how very informal the actual naming process is for Dubois. He simply starts to use the name in the middle of the first page of the report, with no indication that is new, no formal diagnosis or comparison to other taxa.

Paleontological Investigations on Java

Eugène Dubois

The excavations continued during this quarter at *Trinil* yielded, among the remains of more rare vertebrates, a cranium and half a mandible of the pig of type *Sus verrycosus*, otherwise known only from loose molars, a mandible of the smaller variety, very similar to *Sus celebensis*, a large portion of a cranium of *Boselaphus* and another, smaller one of *Garialis*, but the most important find in the month of August was the left thigh bone of *Anthropopithecus*, the existence of which was proven one year ago by a molar and the skullcap. This thigh bone lay at the same level where the other two parts were found, but following the direction of the earlier stream that had deposited the material from the tuff, 15 m upstream. From this discovery and from comparative examination it appears that the three parts of the skeleton belonged to the same individual, probably a female, and surely quite advanced in age.

From Eugène Dubois, "Palaeontologische onderzoekingen op Java," *Verslag van het Mijnwezen*, Third Quarter (1892), pp. 10–14. Translated for use in this volume by the Berkeley Scientific Translation Service.

This discovery brought to light a fact that is as surprising as it is important. The Javanese *Anthropopithecus*, whose cranium also surpasses the highest anthropoids known thus far, had already fully adopted the upright posture, which has long been deemed an exclusive prerogative of the human. Thus, in this Early Pleistocene anthropoid from our island, the first transitional form was found that brings man closer, in unmistakable fashion, to his closest living relatives among the mammals.

By each of the three skeleton pieces found, *Anthropopithecus erectus* Eug. Dubois is closer to man than any other anthropoid, but especially by the thigh bone—a fact that is in perfect agreement with the opinion, first proclaimed by Lamarck and later set forth by Darwin and others, that the first step on the way to our ancestors becoming human must have been achievement of the upright posture.

The skullcap, which by its general form, like the other parts, belongs to the genus *Anthropopithecus* (but also to *Hylobates*), differs in its unusual size—the length was 185 mm, the width, measured in the transverse plane of this line at 1/3 of its length from behind, 130 mm—and also by the strong curvature and the considerably lesser development of the eyebrow ridges. In both of the latter respects, it has the same relationships as were found in *A. troglodytes* at a time of life that can be compared to a seven-year-old human child and in which the cerebral section of the cranium, which grows the least in all the apes, is still relatively much more prominent than is the case in the fully grown animal. One can calculate the cranial capacity, by approximation, to be at least 2.4 times the average of the chimpanzee; and while the brains of the gorilla, which has the largest brain among the anthropoid apes, only attain an average of 1/3 that of the average brain contents of the human, that of this *erectus* (whose body very likely was also built like that of man and had the same size as the average of the European races) must have comprised 2/3 of the cranial capacity of the human.

Of the third upper true molar, the two rear cusps of the crown are even more reduced than is the case with *A. troglodytes* and also with *A. sivalensis*. This reduction is as strong as is that generally of the upper wisdom tooth in man, and the molar has only two roots; however, contrary to the rule in man, the posterior lateral cusp of the crown is less developed here than is the medial cusp. In this respect, the Javanese form agrees with the other two anthropopitheci.

The thigh bone reveals, in dimensions and shape, a striking resemblance to this supporting bone in the human body; it differs therefrom only in less important peculiarities. It is as long and as slender, thus differing greatly from the femur of the large living anthropoids, and the least among these from that of the chimpanzee. Its length is 455 mm, and the ratio between this and the thickness of the shaft is the same as that in a normal adult human, 16-1/2:1 on average. This is of great importance. For since this ratio determines the carrying capacity of the leg and it is known that the carrying capacity here must be in perfect harmony with the load supported, it follows that the upper body was not heavier than that of man and, furthermore—since in the anthropoids as well as in man the thigh length is the same proportion as the length of the entire lower body—that the proportion between the latter and the upper body was a human one and thus entirely different from that of the large living anthropoids, whose bodily relationships differ primarily from that of man in having legs that are very short as compared to the upper body. This creature was therefore in no way adapted to climbing trees in the manner of the chimpanzee, the gorilla, and the orangutan, which have short legs carrying prehensile feet and long arms with a very heavy and long torso for this. On the contrary, it appears from the entire structure of the femur that this leg fulfilled the same mechanical role as in the human body. The head of the joint has the same form and the same radius of curvature as that of a human femur of equal length, and the neck is just as long and forms the same angle with the middle piece as does that of the human. The lower end is just as wide at the condyles and—what is of most importance—the middle of the joint surface at the articulation with the shin bone is directed downward in the same way, and not slanting backwards, as with the anthropoids. In its dimensions, radii of curvature, and forms, this joint surface, like that for the kneecap, agrees in all particulars with that of the human. The angle between the knee base and the anatomical axis and that between the knee base and the mechanical axis are also as large as in man: the position of the thigh bone with respect to the vertical was also slanted and therefore the hips were not less broad. Moreover, the torsion of the femur is just as large. The acute line of this leg is well developed, and the two rotary joints are not different from those parts in man.

Points of difference from man of less importance are the more round shape of the shaft of the thigh bone at the inner side, the lesser development of the lowest part of the oblique line at the front side, the more concave form of the midrotary crest, and the slighter development of the surface of the knee pit. In these respects, the leg conforms to that of the living anthropoid species.

It follows with complete certainty from this examination of the thigh bone that the Javanese *Anthropopithecus* stood and walked in the same upright position as man, which is further confirmed by the similar development and differentiation of the rough spot serving for insertion of the gluteus, since the strong development and high insertion that this muscle has in man, contrary to all lower mammals and apes, is connected with his upright bearing.

Thus, while *A. erectus* used its legs exclusively for locomotion, it must be assumed that, given this division of labor between the front and the hind limbs, the hand was already much more perfected than that organ had become in the anthropoids. For once the hand became readily available, the development of that most perfect and most universal of all tools, that most trustworthy sense, must have made very quick progress, as did the brain, partly as cause, partly as effect thereof. Consistent with this is the human proportion that the upper body seems to have possessed, as well as the higher development of the cranium and the reduction of the teeth. Thus, this creature had the need, and the ability, to seek food elsewhere and in different fashion than by climbing in trees, and to make use of other—artificial—weapons for defense, than its teeth.

One may assume that the differentiation of the organic form will more quickly attain a higher perfection as it is a greater advantage in the struggle for existence under the given circumstances. The peculiar differentiation of the basic anthropoid form, whose starting point lies in the division of labor between the upper and lower limbs, in fact seems to have been highly successful, and there can be no doubt that, once the first and most important step was taken, the further development must have progressed very fast. While we already find certain traces, during the last interglacial period (in Europe), of the existence of man at a level of development not noticeably lower than that of the present day, it also appears quite possible that man has developed from this Early Pleistocene *Anthropopithecus erectus*.

And thus the factual proof is provided of what some have already conjectured, that the East Indies was the cradle of the human kind.

3

Homo heidelbergensis Schoetensack, 1908
Type specimen: Mauer 1

    This selection is a brief excerpt from Otto Schoetensack's extensive monograph describing the Mauer mandible. The holotype (and only specimen) of *Homo heidelbergensis* had been found in Germany just the year before this detailed descriptive and comparative paper was published. The author first describes the geological setting of the discovery and the other fossil fauna recovered from the site. The section of the monograph dealing with the specimen itself consists of accounts of its anatomy and comparisons with the mandibles of other fossil hominids and recent primates. Schoetensack also describes the individual teeth in detail and compares them with other specimens.

    At the time of discovery, it was clear from the associated fauna that the Mauer specimen was considerably older geologically than any Neandertal site and thus the oldest known fossil hominid from Europe. This site has never been precisely dated, but it is still generally considered to be among the oldest in Europe, probably over 500,000 years.

    Despite the many explicit comparisons to other fossils in this monograph, there is no formal taxonomic section. The author simply uses the new name *Homo heidelbergensis* in the title and text, without further taxonomic justification or discussion. (See selections 1 and 2 for similarly casual treatment of the naming process.) In recent decades this specimen has usually been considered either an early member of *Homo sapiens* or a member of *Homo erectus*. See Tattersall (selection 19) for a dissenting view that *Homo heidelbergensis* may represent a distinct species taxon.

The Mandible of *Homo heidelbergensis* from the Mauer Sands at Heidelberg

Otto Schoetensack

The village of Mauer, where we made our find on October 21, 1907, is 10 km southeast of Heidelberg and 6 km south of Neckargemünd, situated near the southern edge of the Odenwald mountains.

. . .

The mammal fauna from the Mauer sands shows close connection to the one from the Mosbach sands. In turn, both reveal clear connections to the pre-glacial Forest Beds of Norfolk as well as to the southern European Upper Pliocene. *Rhinoceros etruscus* Falconer and the horse from Mauer (intermediate between the form *Equus stenonis* Cocchi and the Taubach form) especially indicate a specifically Pliocene age, whereas the rest of the Mammalia for the most part pertain to the oldest Diluvium. The lower jaw from Mauer, therefore, is likely to be the oldest of the stratigraphically confirmed human remains hitherto discovered.

. . .

Now we turn to the description of the lower jaw, overwhelmed from the first view by the individuality of our object. It shows a combination of characteristics hitherto found in neither a recent nor a fossil human mandible. Even an expert could not be blamed if he hesitated to accept it as human. It lacks completely the one

From Otto Schoetensack, *Der Unterkiefer des* Homo heidelbergensis *aus den Sanden von Mauer bei Heidelberg*, pp. 1, 19–20, 25–26, 44–45 (1908). Leipzig: Wilhelm Engelmann. Excerpts translated by W. E. Meikle.

characteristic which is truly human, namely an external protrusion of the chin region; in addition, it combines surprising external dimensions of the body and rami of the lower jaw.

If only a fragment without teeth had been found, it would not have been possible to diagnose this as human. On the basis of a section of the symphysis region one would, with good grounds, assign it to an anthropoid something like a gorilla; looking at a broken part of the ascending ramus, one would think of a large variety of gibbon.

The absolutely positive proof that we are dealing with a human element lies entirely in the nature of the dentition. The completely preserved teeth bear the stamp "human" because the canines show no trace of a strong projection beyond the opposing set of teeth. This is altogether the moderate and harmonic formation with which recent humanity is endowed.

Neither in their dimensions do the teeth of the Heidelberg mandible exceed the range of variation of recent humans. True, their measurements are relatively large compared to modern European examples. However, the difference disappears when extended to recent lower races. In the individual measurements, on the contrary, the teeth—but not the jaw—of *Homo heidelbergensis* would be exceeded by many recent Australians.

. . .

From the comparison of the mandible of *Homo heidelbergensis* with the other discussed fossil jaws it follows that none of these can match our object with regard to morphological significance. The Heidelberg fossil exceeds all of them in its combination of primitive characteristics. The lower jaw from Spy stands relatively closest to it; this appears however proportionately transformed in all elements from the Heidelberg type. The individual variations from Krapina exhibit unique evolutionary paths (perhaps inherited from ancient races).

That the lower jaw of modern races themselves can also be traced back to a prototype quite close to the Heidelberg type was evident previously from several profile diagrams.

Now that the morphological position of our fossil has been illuminated from different orientations, a summary of the results may follow here: The mandible of *Homo heidelbergensis* lets us glimpse the primitive state which the common ancestors of humans and anthropoid apes approached. This find signifies the farthest downward advance in the morphogenesis of the human skeleton recorded to date. Assuming that a geologically still older lower jaw from the ancestral lineage of humans is found, it is unlikely that it would look much different from our fossil, which already brings us to that limit where it requires special evidence (as here of the dentition), to prove membership among the humans. Still further downward we come to the collective common ancestors of the primates. Such a lower jaw would hardly be considered the ancestral stem to modern humans; its connection to our fossil would however certainly be recognizable. This follows from the similarities to it (now in this point, now in that one) found in the mandible of monkeys and recent, as well as fossil, prosimians. In this respect the ramus mandibulae is especially instructive. For example: the resemblance of the processus coronoideus and of the flat incisura semilunaris to *Cynocephalus*; the indication of an incisura subcoronoidea to *Mycetes*; the breadth of the horizontal rami to fossil lemurs.

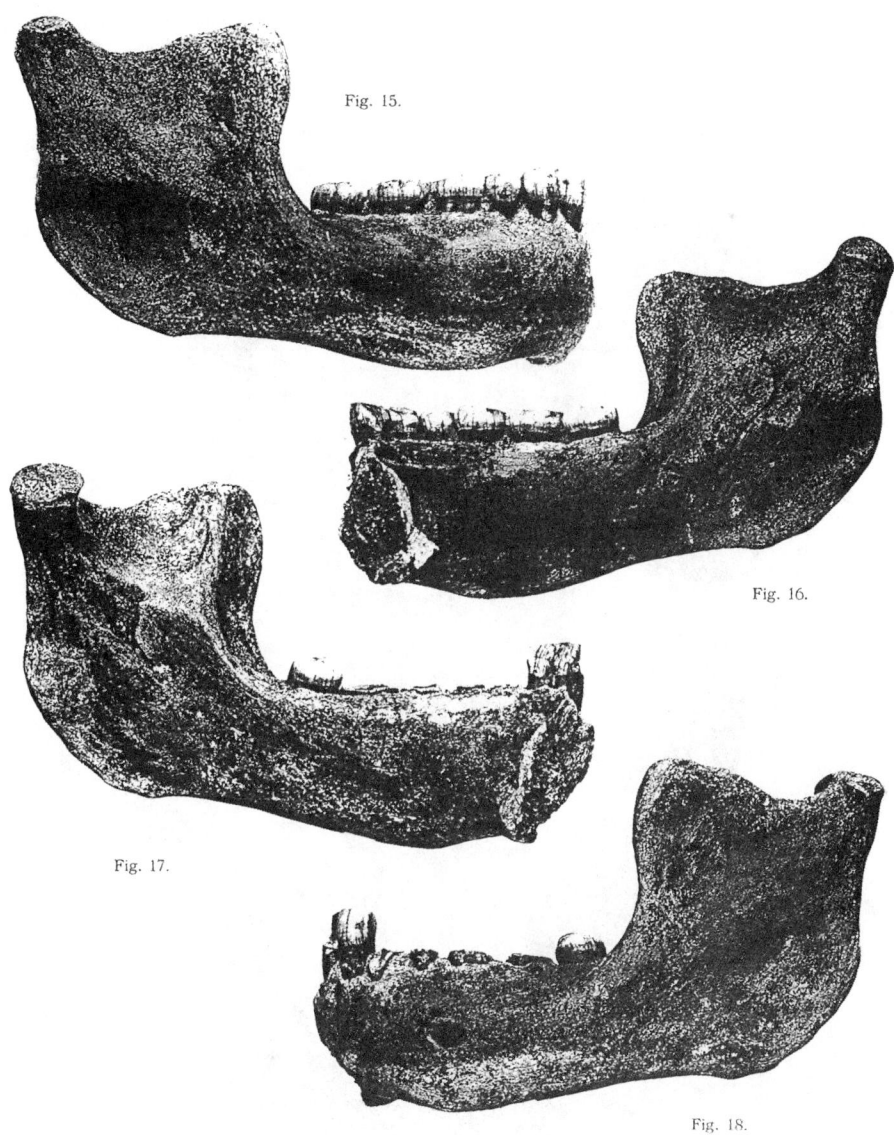

Fig. 15.

Fig. 16.

Fig. 17.

Fig. 18.

Figures 15 and 16. The right half of the mandible of *Homo heidelbergensis* in lateral and medial view.

Figures 17 and 18.   The left half of the mandible in medial and lateral view.

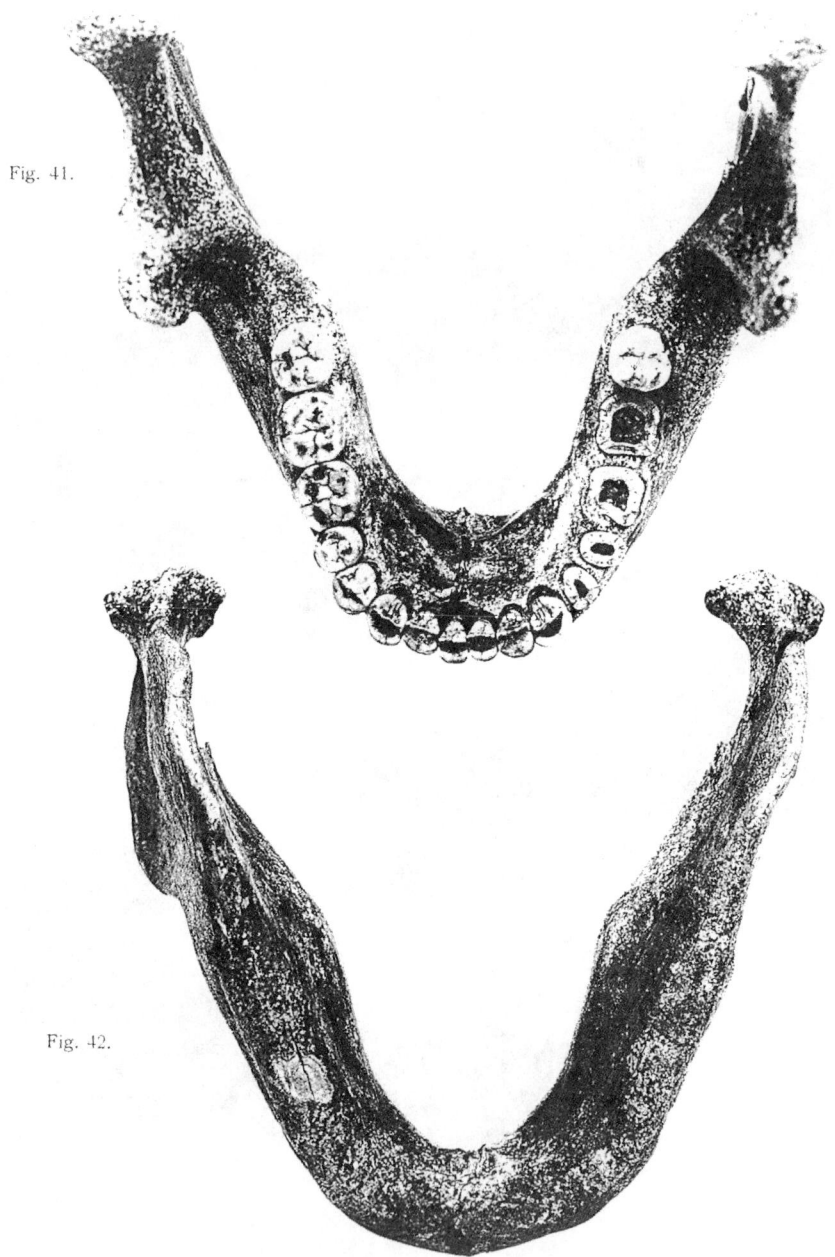

Fig. 41.

Fig. 42.

Figures 41 and 42.   The mandible of *Homo heidelbergensis* seen from above and below in illustration of the dental arch.

4

Homo rhodesiensis Woodward, 1921
Type specimen: Kabwe 1

This paper by Arthur Smith Woodward announces the discovery of a complete skull and some other bones, which came to be known as "Rhodesian Man," from a cave at Broken Hill in what is now Zambia. The site is generally referred to as Kabwe today. Note how Woodward compares the skull especially to Neandertal specimens. Although the context of discovery was not certain, some fauna from the cave led Woodward to suggest a recent age, perhaps post-Pleistocene, and an intermediate phylogenetic position between Neandertals and living humans. The correct geological age of this specimen has never been definitely determined, but today it is considered much older, probably well over 100,000 years.

For some decades following its description, the name *Homo rhodesiensis* was commonly used for this specimen. In recent decades many have placed the Kabwe skull (and other, apparently related specimens from Africa) in the rather amorphous "archaic *Homo sapiens*" group. (See selection 19 for comments by I. Tattersall on the "archaic *Homo sapiens*" concept.) Following the strong trend since World War II to recognize fewer taxa and fewer names among fossil hominids, most anthropologists have retained the name *Homo rhodesiensis* as a sub-species of *H. sapiens* or dropped it entirely. For those who continue to recognize a number of hominid species taxa in the Middle and Upper Pleistocene, beyond the "standard" *Homo erectus* and *Homo sapiens*, however, *Homo rhodesiensis* remains an available name, along with others contained in this collection such as *Homo neanderthalensis* and *Homo heidelbergensis*.

Woodward's paper contains the first scientific announcement of the Kabwe specimen, the first preliminary description of it, and the first proposal of a new species and name with it as type specimen. It is perhaps a little curious that Woodward does not mention Piltdown in his comparisons, given his close association with that "discovery" of less than a decade before.

47

A New Cave Man from Rhodesia, South Africa

Arthur Smith Woodward

During recent years the British Museum has received from the Rhodesia Broken Hill Development Co. numerous bones from a cave discovered in their mine in North-west Rhodesia about 150 miles north of the Kafue river. All except the smaller of these bones are merely broken fragments, and they evidently represent the food of men and flesh-eating mammals who have at different times occupied the cave. As described by Mr. Franklin White (Proc. Rhodesia Sci. Assoc., vol. 7, p. 13, 1908) and Mr. F. P. Mennell (*Geological Magazine* [5], vol. 4, p. 443, 1907), rude stone and bone implements are abundant among the remains, and there can be no doubt that the cave was a human habitation for a long period. Very few of the bones can be exactly named, but, so far as they have been identified by Dr. C. W. Andrews and Mr. E. C. Chubb, they belong to species still living in Rhodesia or to others only slightly different from these. The occupation of the cave, therefore, seems to have been at no distant date—it may not even have been so remote as the Pleistocene period.

Until lately no remains of the cave man himself have been noticed at Broken Hill, but at the end of last summer Mr. W. E. Barren was so fortunate as to discover and dig out of the earth in a remote part of the cave a nearly complete human skull, a fragment of the upper jaw of another, a sacrum, a tibia, and the two ends of a femur. These specimens have just been brought to England by Mr. Ross Macartney, the managing director of the company, and they are to be added to the many generous gifts of the company to the British Museum.

The skull is in a remarkably fresh state of preservation, the bone having merely lost its animal matter and not having been in the

Reprinted by permission from *Nature*, Vol. 108, pp. 371–72. Copyright 1921, Macmillan Magazines, Ltd.

least mineralised. As shown in the accompanying photograph, it is strangely similar to the skull of the Neanderthal or Mousterian race found in the caves of Belgium, France, and Gibraltar. Its brain-case is typically human, with a wall no thicker than that of the average European, and its capacity, though still not determined, is obviously well above the lower human limit. Its large and heavy face is even more simian in appearance than that of Neanderthal man, the great inflated brow-ridges being especially prominent and prolonged to a greater extent at the lateral angles.

The roof of the skull at first sight appears remarkably similar to that of Pithecanthropus from Java, having the same slight median longitudinal ridge along the frontals and rising to its greatest height just about the coronal suture. It is, however, very much larger, and the resemblance may not imply any close affinity. The length of the skull from the middle of the glabella to the inion is about 210 mm., while its maximum width at the parietal bosses is 145 mm. The skull is therefore dolichocephalic, with a cephalic index of 69. Its greatest height (measured from the basion to the bregma) is 131 mm. In general shape the brain-case is much more ordinarily human than that of the La Chapelle Neanderthal skull, which differs in the expansion and bun-shaped depression of its hinder region. The mastoid process, though human, is comparatively small. The supramastoid ridge is very prominent and broad. The tympanic meatus is short and broad, as always in man. The foramen magnum occupies its usual forward position, so that the skull would be perfectly poised on an erect trunk.

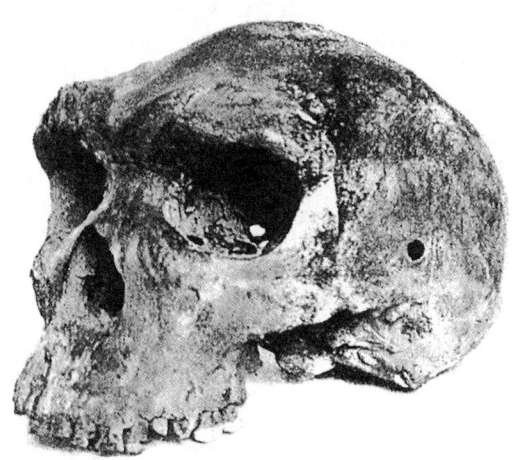

The facial bones much resemble those of the La Chapelle skull, the great flat maxillaries, without canine fossæ, being especially similar. The nasal bones, however, are more gently sloping; the sharp lateral edge of the narial opening runs down on the face (as in the gorilla), allowing the premaxillary surface to pass uninterruptedly into the floor of the narial cavity; and the infranasal region is unusually deep. The typically human anterior nasal spine is conspicuous.

The palate is of enormous size, as large as that inferred by Boule from the fragments preserved in the La Chapelle skull. It is, however, in all respects human, being deeply arched and bounded by the horse-shoe-shaped row of teeth, which are unusually large, but also entirely human. The teeth are much worn, and those of the front of the jaw met their lower opposing teeth in the primitive way, edge to edge. The canines are not enlarged. The second molar is square, 13.5 mm. in diameter. The third molar is much reduced, measuring 12.5 mm. in width by 9.5 mm. in length. The total length of the molar series is about 33 mm. The outside measurement of the dentition across the second molars is 78 mm. The width between the sockets of the third molars is 51 mm. The length from the socket of the median incisor to a line drawn across the back of the third molars is also 51 mm. The whole dentition is much affected with caries, and the disease has spread to the tooth-sockets, which are pierced in some places.

The lower jaw is unfortunately absent, but the size of the palate and the extent of the temporal fossae show that it must have been massive. Even the Heidelberg jaw is slightly narrower and shorter than this must have been.

Although the new skull from the Rhodesian cave so much resembles that of Neanderthal man, the shape of the brain-case and the position of the foramen magnum are so different that we may hesitate to refer the two skulls to the same race. This hesitation seems to be justified when the associated limb-bones are considered, for the tibia is long and slender, of the typically modern type, and the extremities of the femur do not differ in any essential respect from the corresponding parts of a tall and robust modern man. They are thus very different from the tibia and femur of Neanderthal man found in the caves of Belgium and France. As the skull appears to postulate an erect attitude, the congruous limb-bones may well be referred to it. We therefore recognise in the Rhodesian cave man a new form which may be regarded as specifically distinct from *Homo neanderthalensis*, and may be appropriately named *Homo rhodesiensis*.

The precise systematic position of this new species of primitive man can be determined only by further discoveries. It has, however, been pointed out by Prof. Elliot Smith that the refinement of the face was probably the last step in the evolution of the human frame. The newly discovered Rhodesian man may therefore revive the idea that Neanderthal man is truly an ancestor of *Homo sapiens*; for *Homo rhodesiensis* retains an almost Neanderthal face in association with a more modern brain-case and an up-to-date skeleton. He may prove to be the next grade after Neanderthal in the ascending series.

5

Australopithecus africanus Dart, 1925
Type specimen: Taung 1

The Taung discovery is one of the three most significant finds in the history of paleoanthropology (along with those from Neandertal and Trinil) because of its status as the first recognized member of a totally new, previously unknown, major group of fossil hominids. Each of these finds opened up large, new areas of the human fossil record to investigation. This view of Taung's importance is universal today, since it was the first of the Plio-Pleistocene australopithecine specimens and led, via the work of Raymond Dart, Robert Broom, and subsequently many others, to all the rest of the australopithecine finds in South Africa and later in East Africa. Taung's significance is clear in retrospect, given the hundreds of later fossil finds. In the 1920s and 1930s, however, it was completely overshadowed by discoveries of *Homo erectus* at Zhoukoudian, China. Dart's *Nature* paper combined the first announcement of the specimen with taxonomic analysis, preliminary anatomical description, and phylogenetic speculation. The story of the Taung discovery and its aftermath have been told many times and will not be repeated here. For discussion of the history and rôle of Taung, see Tobias (1984).

Raymond Dart's *Australopithecus* stands out as the first new generic name in paleoanthropology which was eventually generally adopted and still continues in use today. All the new genera of fossil hominids proposed before *Australopithecus* (and many after it) are widely considered to be synonyms of *Homo* (such as *Palaeanthropus*, *Pithecanthropus*, *Sinanthropus*, and *Meganthropus*), and are all but unused in the modern literature. (For an exception, see selection 15.)

The Taung specimen is an especially important one in hominid nomenclature. It is the holotype of the species *Australopithecus africanus* Dart, 1925. This species, in turn, is the type species of the genus *Australopithecus* Dart, 1925. This is the oldest available name for any of the australopithecines. It has generally been accepted since the 1950s that the fossils from Sterkfontein and Makapansgat (at least in large part) represent the same species as that from Taung. In contrast, the fossil hominids from Swartkrans and Kromdraai have been viewed almost universally as belonging to a different species, if not another genus, from that at Taung. The suggestion by Tobias (1973) that the Taung individual might actually represent a member of a "robust" rather than "gracile" taxon of South African australopithecine would therefore have had severe nomenclatural consequences if it had been accepted as true. Unless the International Commission on Zoological Nomenclature had acted to remedy the situation, the name *Australopithecus africanus* Dart, 1925 would inevitably have become the correct designation for the "robusts", producing massive confusion in the literature. In addition, the species formerly known as *A. africanus* would have to have taken the next available name; depending on which specimens were included this would likely have been *Australopithecus transvaalensis* Broom, 1936. These points are discussed by Olson (1974) and Tobias (1974; 1978). Luckily for the stability of hominid nomenclature, it does not now seem likely that Taung belongs anywhere other than its traditional position: the first of the "gracile" South African australopithecines to be found (Grine, 1985; Tobias, 1991).

Australopithecus africanus: The Man-Ape of South Africa

Raymond A. Dart

Towards the close of 1924, Miss Josephine Salmons, student demonstrator of anatomy in the University of the Witwatersrand, brought to me the fossilised skull of a cercopithecid monkey which, through her instrumentality, was very generously loaned to the Department for description by its owner, Mr. E. G. Izod, of the Rand Mines Limited. I learned that this valuable fossil had been blasted out of the limestone cliff formation—at a vertical depth of 50 feet and a horizontal depth of 200 feet—at Taungs, which lies 80 miles

Reprinted by permission from *Nature*, Vol. 115, pp. 195–99. Copyright 1925 Macmillan Magazines, Ltd.

north of Kimberley on the main line to Rhodesia, in Bechuanaland, by operatives of the Northern Lime Company. Important stratigraphical evidence has been forthcoming recently from this district concerning the succession of stone ages in South Africa (Neville Jones, *Jour. Roy. Anthrop. Inst.*, 1920), and the feeling was entertained that this lime deposit, like that of Broken Hill in Rhodesia, might contain fossil remains of primitive man.

I immediately consulted Dr. R. B. Young, professor of geology in the University of the Witwatersrand, about the discovery, and he, by a fortunate coincidence, was called down to Taungs almost synchronously to investigate geologically the lime deposits of an adjacent farm. During his visit to Taungs, Prof. Young was enabled, through the courtesy of Mr. A. F. Campbell, general manager of the Northern Lime Company, to inspect the site of the discovery and to select further samples of fossil material for me from the same formation. These included a natural cercopithecid endocranial cast, a second and larger cast, and some rock fragments disclosing portions of bone. Finally, Dr. Gordon D. Laing, senior lecturer in anatomy, obtained news, through his friend Mr. Ridley Hendry, of another primate skull from the same cliff. This cercopithecid skull, the possession of Mr. De Wet, of the Langlaagte Deep Mine, has also been liberally entrusted by him to the Department for scientific investigation.

The cercopithecid remains placed at our disposal certainly represent more than one species of catarrhine ape. The discovery of Cercopithecidæ in this area is not novel, for I have been informed that Mr. S. Haughton has in the press a paper discussing at least one species of baboon from this same spot (Royal Society of South Africa). It is of importance that, outside of the famous Fayüm area, primate deposits have been found on the African mainland at Oldaway (Hans Reck, *Sitzungsbericht der Gesellsch. Naturforsch. Freunde*, 1914), on the shores of Victoria Nyanza (C. W. Andrews, *Ann. Mag. Nat. Hist.*, 1916), and in Bechuanaland, for these discoveries lend promise to the expectation that a tolerably complete story of higher primate evolution in Africa will yet be wrested from our rocks.

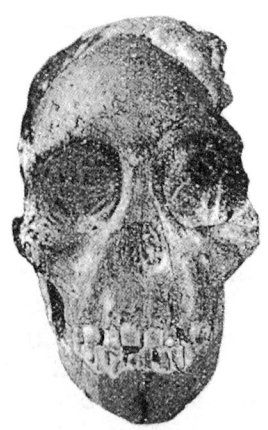

Figure 1. Norma facialis of *Australopithecus africanus* aligned on the Frankfort horizontal.

In manipulating the pieces of rock brought back by Prof. Young, I found that the larger natural endocranial cast articulated exactly by its fractured frontal extremity with another piece of rock in which the broken lower and posterior margin of the left side of a mandible was visible. After cleaning the rock mass, the outline of the hinder and lower part of the facial skeleton came into view. Careful development of the solid limestone in which it was embedded finally revealed the almost entire face depicted in the accompanying photographs.

It was apparent when the larger endocranial cast was first observed that it was specially important, for its size and sulcal pattern revealed sufficient similarity with those of the chimpanzee and gorilla to demonstrate that one was handling in this instance an anthropoid and not a cercopithecid ape. Fossil anthropoids have not hitherto been recorded south of the Fayüm in Egypt, and living anthropoids have not been discovered in recent times south of Lake Kivu region in Belgian Congo, nearly 2,000 miles to the north, as the crow flies.

All fossil anthropoids found hitherto have been known only from mandibular or maxillary fragments, so far as crania are concerned, and so the general appearance of the types they represented has been unknown; consequently, a condition of affairs where virtually the whole face and lower jaw, replete with teeth, together with the major portion of the brain pattern, have been preserved, constitutes a specimen of unusual value in fossil anthropoid discovery. Here, as in *Homo rhodesiensis*, Southern Africa has provided documents of higher primate evolution that are amongst the most complete extant.

Apart from this evidential completeness, the specimen is of importance because it exhibits an extinct race of apes *intermediate between living anthropoids and man.*

In the first place, the whole cranium displays *humanoid* rather than anthropoid lineaments. It is markedly dolichocephalic and leptoprosopic, and manifests in a striking degree the *harmonious relation* of calvaria to face emphasised by Pruner-Bey. As Topinard says, "A cranium elongated from before backwards, and at the same time elevated, is already in harmony by itself; but if the face, on the other hand, is elongated from above downwards, and narrows, the harmony is complete." I have assessed roughly the difference in the relationship of the glabella-gnathion facial length to the glabella-inion calvarial length in recent African anthropoids of an age comparable with that of this specimen (depicted in Duckworth's "Anthropology and Morphology," second edition, vol. i.), and find that, if the glabella-inion length be regarded in all three as 100, then the glabella-gnathion length in the young chimpanzee is approximately 88, in the young gorilla 80, and in this fossil 70, which proportion suitably demonstrates the enhanced relationship of cerebral length to facial length in the fossil (Figure 2).

The glabella is tolerably pronounced, but any traces of the salient supra-orbital ridges, which are present even in immature living anthropoids, are here entirely absent. Thus the relatively increased glabella-inion measurement is due to brain and not to bone. Allowing 4 mm. for the bone thickness in the inion region, that measurement in the fossil is 127 mm.; *i.e.* 4 mm. less than the same measurement in an adult chimpanzee in the Anatomy Museum at the University of the Witwatersrand. The orbits are not in any sense detached from the forehead, which rises steadily from their margins in a fashion amazingly human. The interorbital width is very small (13 mm.) and the ethmoids are not blown out laterally as in modern African anthropoids. This lack of ethmoidal expansion causes the lacrimal fossæ to face posteriorly and to lie relatively far back in the orbits, as in man. The orbits, instead of being subquadrate as in anthropoids, are almost circular, furnishing an orbital index of 100, which is well within the range of human variation (Topinard, "Anthropology"). The malars, zygomatic arches, maxillæ, and mandible all betray a delicate and humanoid character. The facial prognathism is relatively slight, the gnathic index of Flower giving a value of 109, which is scarcely greater than that of certain Bushmen (Strandloopers) examined by Shrubsall. The nasal bones are not prolonged below the level of the lower orbital margins, as in anthropoids, but end above these, as in man, and are incompletely fused together in their lower half. Their maximum length (17 mm.) is not so great as that of the nasals in *Eoanthropus dawsoni*. They are depressed in the median line, as in the chimpanzee, in their lower half, but it seems probable that this depression has occurred post-mortem, for the upper half of each bone is arched forwards (Figure 1). The nasal aperture is small and is just wider than it is high (17 mm. x 16 mm.). There is no nasal spine, the floor of the nasal cavity being continuous with the anterior aspect of the alveolar portions of the maxillæ, after the fashion of the chimpanzee and of certain New Caledonians and negroes (Topinard, *loc. cit.*).

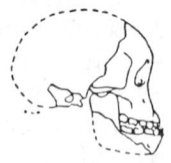

AUSTRALOPITHECUS

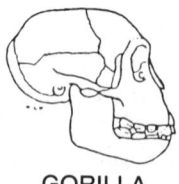

GORILLA

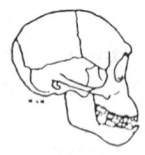

CHIMPANZEE

Figure 2.   Cranial form in living anthropoids of similar age (after Duckworth) and in the new fossil. For this comparison, the fossil is regarded as having the same calvarial length as the gorilla.

In the second place, the dentition is *humanoid* rather than anthropoid. The specimen is juvenile, for the first permanent molar tooth only has erupted in both jaws on both sides of the face; *i.e.* it corresponds anatomically with a human child of six years of age. Observations upon the milk dentition of living primates are few, and only one molar tooth of the deciduous dentition in one fossil anthropoid is known (Gregory, "The Origin and Evolution of the Human Dentition," 1920). Hence the data for the necessary comparisons are meagre, but certain striking features of the milk dentition of this creature may be mentioned. The tips of the canine teeth transgress very slightly (0.5–0.75 mm.) the general margin of the teeth in each jaw, *i.e.* very little more than does the human milk canine. There is no diastema whatever between the premolars and canines on either side of the lower jaw, such as is present in the deciduous dentition of living anthropoids; but the canines in this jaw come, as in the human jaw, into alignment with the incisors (Gregory, *loc. cit.*). There is a diastema (2 mm. on the right side, and 3 mm. on the left side) between the canines and lateral incisors of the upper jaw; but seeing, first, that the incisors are narrow, and, secondly, that diastemata (1 mm.–1.5 mm.) occur between the central incisors of the upper jaw and between the medial and lateral incisors of both sides in the lower jaw, and, thirdly, that some separation of the milk teeth takes place even in mankind (Tomes, "Dental Anatomy," seventh edition) during the establishment of the permanent dentition, it is evident that the diastemata which occur in the upper jaw are small. The lower canines, nevertheless, show wearing facets both for the upper canines and for the upper lateral incisors.

The incisors as a group are irregular in size, tend to overlap one another, and are almost vertical, as in man; they are not symmetrical and well spaced, and do not project forwards markedly, as in anthropoids. The upper lateral incisors do project forwards to some extent and perhaps also do the upper central incisors very slightly, but the lateral lower incisors betray no evidence of forward projection, and the central lower incisors are not even vertical as in most races of mankind, but are directed slightly backwards, as *sometimes* occurs in man. Owing to these remarkably human characters displayed by the deciduous dentition, when contour tracings of the upper jaw are made, it is found that the jaw and the teeth, as a whole, take up a parabolic arrangement comparable only with that presented by mankind amongst the higher primates. These facts, together with the more minute anatomy of the teeth, will be illustrated and discussed in the memoir which is in the process of elaboration concerning the fossil remains.

In the third place, the mandible itself is *humanoid* rather than anthropoid. Its ramus is, on the whole, short and slender as com-pared with that of anthropoids, but the bone itself is more massive than that of a human being of the same age. Its symphyseal region is virtually complete and reveals anteriorly a more vertical outline than is found in anthropoids or even in the jaw of Piltdown man. The anterior symphyseal surface is scarcely less vertical than that of Heidelberg man. The posterior symphyseal surface in living anthro-poids differs from that of modern man in possessing a pronounced posterior prolongation of the lower border, which joins together the two halves of the mandible, and so forms the well-known *simian shelf* and above it a deep genial impression for the attachment of the tongue musculature. In this character, *Eoanthropus dawsoni* scarcely differs from the anthropoids, especially the chimpanzee; but this new fossil betrays no evidence of such a shelf, the lower border of the mandible having been massive and rounded after the fashion of the mandible of *Homo heidelbergensis.*

That hominid characters were not restricted to the face in this extinct primate group is borne out by the relatively forward situation of the foramen magnum. The position of the basion can be assessed within a few millimetres of error, because a portion of the right exoccipital is present alongside the cast of the basal aspect of the cerebellum. Its position is such that the basi-prosthion measurement is 89 mm., while the basi-inion measurement is at least 54 mm. This relationship may be expressed in the form of a "head-balancing" index of 60.7. The same index in a baboon provides a value of 41.3, in an adult chimpanzee 50.7, in Rhodesian man 83.7, in a dolichocephalic European 90.9, and in a brachycephalic European 105.8. It is significant that this index, which indicates in a measure the poise of the skull upon the vertebral column, points to the assumption by this fossil group of an attitude appreciably more erect than that of modern anthropoids. The improved poise of the head, and the better posture of the whole body framework which accompanied this alteration in the angle at which its dominant member was supported, is of great significance. It means that a greater reliance was being placed by this group upon the feet as organs of progression, and that the hands were being freed from their more primitive function of accessory organs of locomotion. Bipedal

Figure 3.   Norma lateralis of *Australopithecus africanus* aligned on the Frankfort horizontal.

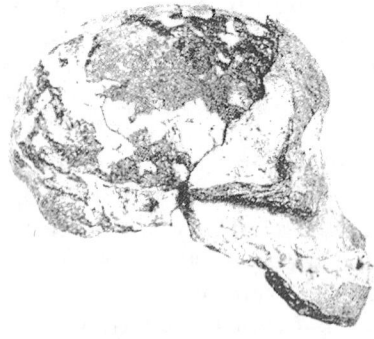

Figure 4. Norma basalis of *Australopithecus africanus* aligned on the Frankfort horizontal.

animals, their hands were assuming a higher evolutionary rôle not only as delicate tactual, examining organs which were adding copiously to the animal's knowledge of its physical environment, but also as instruments of the growing intelligence in carrying out more elaborate, purposeful and skilled movements, and as organs of offence and defence. The latter is rendered the more probable, in view, first, of their failure to develop massive canines and hideous features, and, secondly, of the fact that even living baboons and anthropoid apes can and do use sticks and stones as implements and as weapons of offence ("Descent of Man," p. 81 *et seq.*).

Lastly, there remains a consideration of the endocranial cast which was responsible for the discovery of the face. The cast comprises the right cerebral and cerebellar hemispheres (both of which fortunately meet the median line throughout their entire dorsal length) and the anterior portion of the left cerebral hemisphere. The remainder of the cranial cavity seems to have been empty, for the left face of the cast is clothed with a picturesque lime crystal deposit; the vacuity in the left half of the cranial cavity was probably responsible for the fragmentation of the specimen during the blasting. The cranial capacity of the specimen may best be appreciated by the statement that the length of the cavity could not have been less than 114 mm., which is 3 mm. greater than that of an adult chimpanzee in the Museum of the Anatomy Department in the University of the Witwatersrand, and only 14 mm. less than the greatest length of the cast of the endocranium of a gorilla chosen for casting on account of its great size. Few data are available concerning the expansion of brain matter which takes place in the living anthropoid brain between the time of eruption of the first permanent molars and the time of their becoming adult. So far as man is concerned, Owen ("Anatomy of Vertebrates," vol. iii.) tells us that "The brain has advanced to near its term of size at about ten years, but it does not usually obtain its full development till between twenty and thirty years of age." R. Boyd (1860) discovered an increase in weight of nearly 250 grams in the brains of male human beings after they had reached the age of seven years. It is therefore reasonable to believe that the adult forms typified by our present specimen possessed brains which were larger than that of this juvenile specimen, and equalled, if they did not actually supersede, that of the gorilla in absolute size.

Whatever the total dimensions of the adult brain may have been, there are not lacking evidences that the brain in this group of fossil forms was distinctive in type and was an instrument of greater intelligence than that of living anthropoids. The face of the endocranial cast is scarred unfortunately in several places (cross-hatched in the dioptographic tracing—see figure 5). It is evident that the relative proportion of cerebral to cerebellar matter in this brain was not greater than in the gorilla's. The brain does not show that general pre- and post-Rolandic flattening characteristic of the living anthropoids, but presents a rounded and well-filled-out contour, which points to a symmetrical and balanced development of the faculties of associative memory and intelligent activity. The pithecoid type of parallel sulcus is preserved, but the sulcus lunatus has been thrust backwards towards the occipital pole by a pronounced general bulging of the parieto-temporo-occipital association areas.

To emphasise this matter, I have reproduced (Figure 6) super-imposed coronal contour tracings taken at the widest part of the parietal region in the gorilla endocranial cast and in this fossil. Nothing could illustrate better the mental gap that exists between living anthropoid apes and the group of creatures which the fossil represents than the flattened atrophic appearance of the parietal region of the brain (which lies between the visual field on one hand, and the tactile and auditory fields on the other) in the former and its surgent vertical and dorso-lateral expansion in the latter. The expansion in this area of the brain is the more significant in that

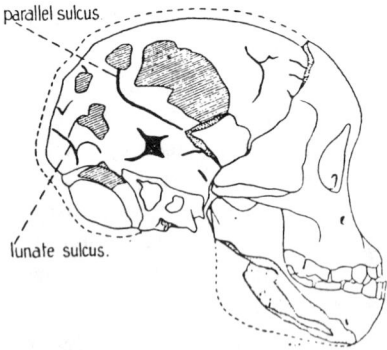

Figure 5.   Dioptographic tracing of *Australopithecus africanus* (right side), x 1/3.

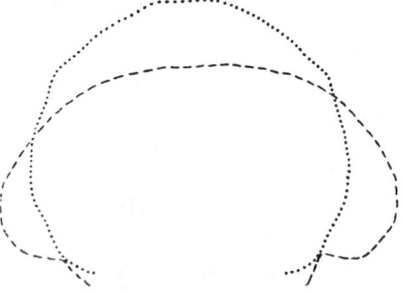

Figure 6.   Contour tracings of coronal sections through the widest part of the parietal region of the endocranial casts in Australopithecus ...... and in a gorilla -------.

it explains the posterior *humanoid* situation of the sulcus lunatus. It indicates (together with the narrow interorbital interval and human characters of the orbit) the fact that this group of beings, having acquired the faculty of stereoscopic vision, had profited beyond living anthropoids by setting aside a relatively much larger area of the cerebral cortex to serve as a storehouse of information concerning their objective environment as its details were simultaneously revealed to the senses of vision and touch, and also of hearing. They possessed to a degree unappreciated by living anthropoids the use of their hands and ears and the consequent faculty of associating with the colour, form, and general appearance of objects, their weight, texture, resilience, and flexibility, as well as the significance of sounds emitted by them. In other words, their eyes saw, their ears heard, and their hands handled objects with greater meaning and to fuller purpose than the corresponding organs in recent apes. They had laid down the foundations of that discriminative knowledge of the appearance, feeling, and sound of things that was a necessary milestone in the acquisition of articulate speech.

There is, therefore, an ultra-simian quality of the brain depicted in this immature endocranial cast which harmonises with the ultra-simian features revealed by the entire cranial topography and corroborates the various inferences drawn therefrom. The two thousand miles of territory which separate this creature from its nearest living anthropoid cousins is indirect testimony to its increased intelligence and mastery of its environment. It is manifest that we are in the presence here of a pre-human stock, neither chimpanzee nor gorilla, which possesses a series of differential characters not encountered hitherto in any anthropoid stock. This complex of characters exhibited is such that it cannot be interpreted as belonging to a form ancestral to any living anthropoid. For this reason, we may be equally confident that there can be no question here of a primitive anthropoid stock such as has been recovered from the Egyptian Fayüm. Fossil anthropoids, varieties of Dryopithecus, have been retrieved in many parts of Europe, Northern Africa, and Northern India, but the present specimen, despite its youth, cannot be confused with anthropoids having the dryopithecid dentition. Other fossil anthropoids from the Siwalik hills in India (Miocene and Pliocene) are known which, according to certain observers, may be ancestral to modern anthropoids and even to man.

Whether our present fossil is to be correlated with the discoveries made in India is not yet apparent; that question can only be solved by a careful comparison of the permanent molar teeth from both localities. It is obvious, meanwhile, that it represents a fossil group distinctly advanced beyond living anthropoids in those two dominantly human characters of facial and dental recession on one hand, and improved quality of the brain on the other. Unlike Pithecanthropus, it does not represent an ape-like man, a caricature of precocious hominid failure, but a creature well advanced beyond modern anthropoids in just those characters, facial and cerebral, which are to be anticipated in an extinct link between man and his simian ancestor. At the same time, it is equally evident that a creature with anthropoid brain capacity, and lacking the distinctive, localised temporal expansions which appear to be concomitant with and necessary to articulate man, is no true man. It is therefore logically regarded as a man-like ape. I propose tentatively, then, that a new family of *Homo-simiadæ* be created for the reception of the group of individuals which it represents, and that the first known species of the group be designated *Australopithecus africanus*, in commemoration, first, of the extreme southern and unexpected horizon of its discovery, and secondly, of the continent in which so many new and important discoveries connected with the early history of man have recently been made, thus vindicating the Darwinian claim that Africa would prove to be the cradle of mankind.

It will appear to many a remarkable fact that an ultra-simian and pre-human stock should be discovered, in the first place, at this extreme southern point in Africa, and, secondly, in Bechuanaland, for one does not associate with the present climatic conditions obtaining on the eastern fringe of the Kalahari desert an environment favourable to higher primate life. It is generally believed by geologists (*vide* A. W. Rogers, "Post-Cretaceous Climates of South Africa," *South African Journal of Science*, vol. xix., 1922) that the climate has fluctuated within exceedingly narrow limits in this country since Cretaceous times. We must therefore conclude that it was only the enhanced cerebral powers possessed by this group which made their existence possible in this untoward environment.

In anticipating the discovery of the true links between the apes and man in tropical countries, there has been a tendency to overlook the fact that, in the luxuriant forests of the tropical belts, Nature was supplying with profligate and lavish hand an easy and sluggish solution, by adaptive specialisation, of the problem of existence in creatures so well equipped mentally as living anthropoids are. For the production of man a different apprenticeship was needed to sharpen the wits and quicken the higher manifestations of intellect—a more open veldt country where competition was keener between swiftness and stealth, and where adroitness of thinking and movement played a preponderating rôle in the preservation of the species. Darwin has said, "no country in the world abounds in a greater degree with dangerous beasts than Southern Africa," and, in my opinion, Southern Africa, by providing a vast open country with occasional wooded belts and a relative scarcity of water, together with a fierce and bitter mammalian competition, furnished a laboratory such as was essential to this penultimate phase of human evolution.

In Southern Africa, where climatic conditions appear to have fluctuated little since Cretaceous times, and where ample dolomitic formations have provided innumerable refuges during life, and burial-places after death, for our troglodytic forefathers, we may confidently anticipate many complementary discoveries concerning this period in our evolution.

In conclusion, I desire to place on record my indebtedness to Miss Salmons, Prof. Young, and Mr. Campbell, without whose aid the discovery would not have been made; to Mr. Len Richardson for providing the photographs; to Dr. Laing and my laboratory staff for their willing assistance; and particularly to Mr. H. Le Helloco, student demonstrator in the Anatomy Department, who has prepared the illustrations for this preliminary statement.

6

Paranthropus robustus Broom, 1938
Type specimen: TM 1517

    This paper was the first announcement of the discovery, first
description, and first naming of a "robust" australopithecine, the
type specimen from Kromdraai of *Paranthropus robustus* Broom,
1938. The designation "robust", of course, comes from the name of
this species. The subjective, taxonomic, question of whether to
employ Robert Broom's generic name *Paranthropus* or whether to
include this species in *Australopithecus* has been long debated and
promises to remain open into the foreseeable future. The majority of
anthropologists probably refer to this species as *Australopithecus
robustus* today, but a growing number of specialists have reverted
to the use of *Paranthropus* in recent years (Grine, 1988a).

    Robert Broom was a prolific namer of taxa in a variety of fossil
vertebrate groups, especially the mammal-like reptiles and the Plio-
Pleistocene mammals. Note that in this paper he performs several
taxonomic acts. He creates the genus *Paranthropus* and the species
Paranthropus robustus; he also transfers his previously named
species from Sterkfontein, *Australopithecus transvaalensis*, to a
new genus, *Plesianthropus*, based on "the shape of the
symphysis." For more about Broom and his work, see Findlay
(1972).

The Pleistocene Anthropoid Apes of South Africa

Robert Broom

Dart's discovery in 1924 of the fossil ape of Taungs, which he named *Australopithecus africanus*, opened a new chapter in the history of the origin of man. The type skull, which unfortunately is the only one known from that locality, is that of a five-year-old child, and though there seems little doubt that Dart was right in regarding it as an ape much nearer to man than either the chimpanzee or the gorilla, some European men of science still seem to believe that it is a variety of chimpanzee or a dwarf gorilla, in spite of the fact that the milk teeth are entirely different in structure from those of the living anthropoids, and closely similar to those of man.

In 1936 I discovered, at Sterkfontein, much of an adult skull which I described as *Australopithecus transvaalensis*. It is clearly allied to the Taungs ape, but there are few points in which a comparison can be made between the two, and I provisionally placed it in the same genus. In the last two years almost continuous exploration has been going on at Sterkfontein, and many interesting further remains have been found, notices of some of which have been published in *Nature*.

Until this year, nothing was known of the lower jaw except a beautifully preserved third molar. We still do not know much of the mandible, but we now have a well-preserved second premolar, much of what I regard as a female canine, and the incisor portion of the jaw of a young male, corresponding to a human boy of nine years, with the perfectly preserved crown of an unworn canine. This canine is unlike that of any ape at present known, but there seems little doubt that it is rightly identified as that of the male *A. transvaalensis*, from the resemblance it bears in a number of respects to the canine, which I regard as the lower canine of the

Reprinted by permission from *Nature*, Vol. 142, pp. 377–79. Copyright 1938 Macmillan Magazines, Ltd.

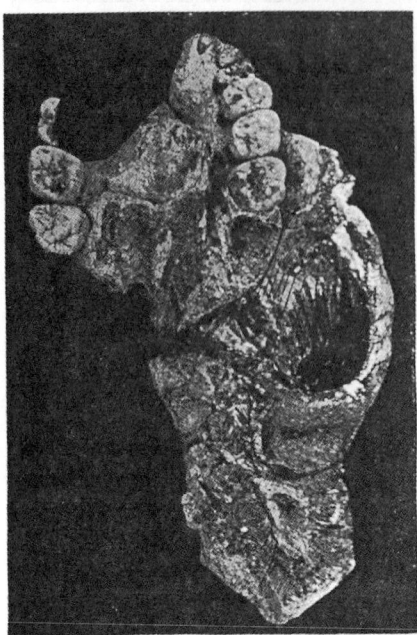

Figure 1.   Palatal view of skull of *Paranthropus robustus* Broom, about ½ natural size. The teeth of the left side have been weathered off, but are replaced in what must have been nearly their original position. Part of the socket of left m^3 is preserved.

female. Though little more than the incisor portion of the symphysis is preserved, it shows the sockets of the incisors, and reveals the interesting fact that the lateral incisors are considerably larger than the central ones. The shape of the symphysis is so different from that of the Taungs ape that it seems advisable to place *A. transvaalensis* in a distinct genus, for which the name *Plesianthropus* is proposed.

In June of this year a most important new discovery was made. A schoolboy, Gert Terblanche, found in an outcrop of bone breccia near the top of a hill, a couple of miles from the Sterkfontein caves, much of the skull and lower jaw of a new type of anthropoid. Not realizing the value of the find, he damaged the specimen considerably in hammering it out of the rock. The palate with one molar tooth he gave to Mr. Barlow at Sterkfontein, from whom I obtained it. Recognizing that some of the teeth had recently been broken off, and that there must be other parts of the skull where the palate was found, I had to hunt up the schoolboy. I went to his home two miles off and found that he was at school another two miles away, and his mother told me that he had four beautiful teeth with him. I naturally went to the school, and found the boy with four of what are perhaps the most valuable teeth in the world in his trouser pocket. He told me that there were more bits of the skull on the hillside. After school he took me to the place and I gathered every scrap I could find; and when these were later examined and cleaned and joined up, I found I had not only the nearly perfect palate with most of the teeth, but also practically the whole of the left side of the lower half of the skull and the nearly complete right lower jaw. The only missing parts of importance are the halves of two molars, the crown of the left 1st upper premolar and the crown

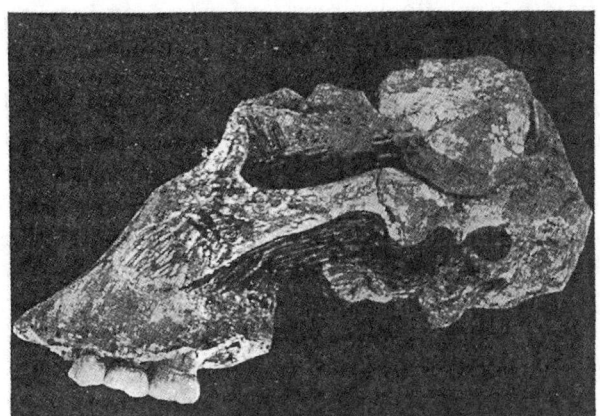

Figure 2.   Side view of skull of *Paranthropus robustus* Broom. About 1/2 natural size.

of the right lower canine. Those I still hope to discover. As, however, we have impressions in the matrix of some of the missing teeth and parts, we know nearly the complete definition.

The skull is that of a large ape, larger than most male chimpanzees and nearly as large as most female gorillas; but it differs very greatly from both the living African anthropoids. Much of the palate is preserved in perfect condition. The whole of the left side of the sphenoid bone is also preserved; while the zygomatic arch is nearly complete. The glenoid cavity and the tympanic bone are in perfect preservation, and much of the mastoid region, and part of the occiput with a portion of the left condyle.

The glenoid cavity and the relations to the tympanic bone are of exceptional interest. In the gorilla, the chimpanzee, the orang and the gibbon, the outer part of the tympanic is situated behind the posterior glenoid process. In man, the tympanic is situated mainly below the glenoid process, and even at its outer part it forms the posterior non-articular part of the glenoid cavity. In the new fossil ape, the condition of the glenoid and tympanic is almost exactly as in man, though the parts are very much larger.

The occipital condyle is in practically the same plane as the external auditory meatus and thus farther forward than in the gorilla and the chimpanzee; which appears to indicate that the ape walked somewhat more erect than the living anthropoids.

From the portion of the brain case preserved, I estimate the volume of the brain to have been about 600 c.c. The face is remarkably flat and much shorter than in the gorilla. A curious bony ridge runs down from the inner border of the large infraorbital foramen.

The molar teeth, as will be seen from the illustrations, differ considerably in shape from those in *Plesianthropus transvaalensis*, and the second premolar is about half as large again as in the Sterkfontein ape. The upper canine had been lost before fossilization, but it must have been relatively remarkably small, and the incisors, of which we have much of the sockets preserved, were also relatively small. The palate is relatively short and broad, and owing to the small size of the incisors and canines the anterior part is narrowed, and the teeth are arranged more as in man than in any of the living anthropoids. The anterior two-thirds of the right mandible are satisfactorily preserved. The symphyseal region has been broken off behind the canine before fossilization and slightly displaced. The incisors which are lost have been relatively very small, and the lateral ones are scarcely larger than the central. The canine crown is lost, but the impression of its outer side is preserved in the matrix. It is quite a small tooth, and remarkably human in shape. It is clearly very unlike the canine of *Plesianthropus transvaalensis*. The premolars have rounded crowns without any high well-developed cusps as in the living anthropoids, and are thus fairly similar to those of man, but about twice as large. The second premolar differs very markedly from that of *Plesianthropus transvaalensis*, and we may thus confidently place the new skull in a new genus and species.

The deposit in which the skull was found is the floor of an old cave the walls of which have probably been weathered away thousands of years ago. We may therefore suspect that the deposit is very much older than that in the Sterkfontein caves, and this is confirmed by the associated fauna. It contains a jackal, a baboon, a horse and a hyrax, which are all of different species from those at Sterkfontein, and are most probably all older. The skull may be referred to as the Kromdraai (pronounced 'Kromdry') skull, and may be given the name *Paranthropus robustus*.

It seems probable that the Sterkfontein skull is of Upper Pleistocene age, the Kromdraai skull of Middle Pleistocene and the Taungs skull probably of Lower Pleistocene; though of course more work will have to be done before the geological ages of any of these skulls can be determined with more than probability.

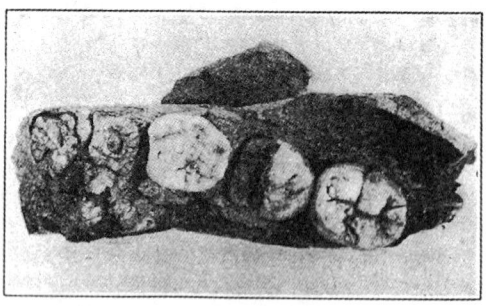

Figure 3.   Occlusal view of molars, with roots of premolars, of right mandible of *Paranthropus robustus* Broom.

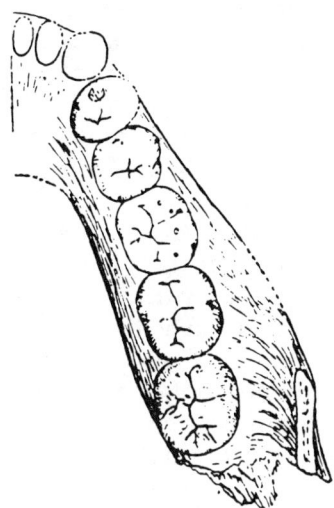

Figure 4.   Occlusal view of right mandible of *Paranthropus robustus* Broom. 3/4 natural size. The portion of the jaw with canine and incisor sockets was detached and is placed in what was probably its actual relationship.

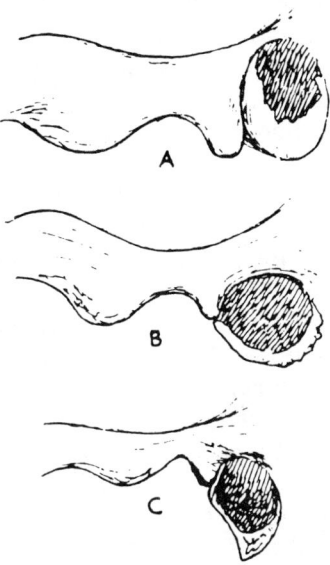

Figure 5.   The glenoid and the external auditory meatus in (*A*) a large male gorilla; (*B*) *Paranthropus robustus*; and (*C*) a large Korana male. 3/4 natural size.

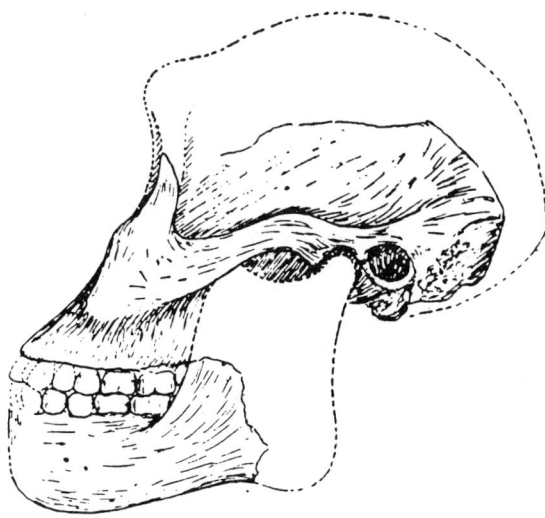

Figure 6.   Side View of skull of *Paranthropus robustus* Broom. 3/8 natural size. Restored. The parts in line are known.

Clearly, during the Pleistocene there lived in South Africa a number of large-brained anthropoids which resemble man in the shape of their premolars and in having relatively small canines, and in having the glenoid region, in at least some forms, remarkably human in structure. These Pleistocene apes are probably the modified descendants of forms that may have been widely distributed over Africa in Pliocene times, and it is probably from one of the Pliocene members of the group that man arose.

Quite certainly the conditions in Pleistocene times in South Africa were not unlike those of today. The apes lived on the plains and among the rocky krantzes. At Taungs most probably the apes lived in the caves. The associated animal remains seem to be the kitchen midden of *Australopithecus*.

At Sterkfontein and Kromdraai the larger bones in the caves seem all to have been introduced by carnivorous animals and the small bones by owls. Nearly every bone of the larger animals has been broken in pieces. No perfect limb bone has been found, and most teeth are detached from the jaws and many of the teeth have also been broken before fossilization.

We have now a fairly good knowledge of the faunas associated with the apes. We know about a dozen fossil mammals from the Taungs caves, all extinct; about thirty mammals from Sterkfontein, nearly all extinct; and we know about a dozen from Kromdraai, all extinct except one—the living porcupine.

7

Paranthropus crassidens Broom, 1949
Type specimen: SK 6

Here Robert Broom announces the first fossil hominid discovery from Swartkrans and gives it the new species name *Paranthropus crassidens*. Typically for Broom, he suggests that the find probably represents a new genus, but he settles for creating just a new species of *Paranthropus*. Australopithecine fossils from Swartkrans have very commonly been grouped into the same species as those from Kromdraai, with the prior name *Paranthropus robustus* (or *Australopithecus robustus*) applied to both. However, a few workers (Grine, 1988b) have continued to recognize a specific distinction, and therefore to retain both of Broom's species names.

Another New Type of Fossil Ape-man

Robert Broom

Some years ago, I pointed out that in my opinion the cave deposits in the dolomite of the Transvaal when fully worked will give us the remains of most stages of early man and pre-man that have

Reprinted by permission from *Nature*, Vol. 163, p. 57. Copyright 1949 Macmillan Magazines, Ltd.

inhabited South Africa from probably Middle Pliocene to Recent. In the Sterkfontein area alone there are apparently in about ten square miles more than a hundred different cave deposits, and many of those are of quite different ages judging by the faunas. Almost all animals at Kromdraai main deposit are quite different species from those at the main quarry at Sterkfontein. The jackals, the sabre-toothed tigers, the baboons, the dassies and the ape-men are quite different species.

At the beginning of November we started work at a new spot in association with the California University Expedition. Though the deposit is on the farm Swartkrans and only a mile from the main Sterkfontein quarry, the fauna so far as we have gone proves to be very different—whether older or younger we cannot yet say. Luckily we found teeth of a new type of ape-man within ten days, and a week later discovered much of a mandible with the complete lower premolars and molars.

The new mandible is not closely allied to that of the Sterkfontein ape-man *Plesianthropus*; the teeth are allied to those of the Kromdraai ape-man *Paranthropus*; but they are much larger and differ in a number of respects.

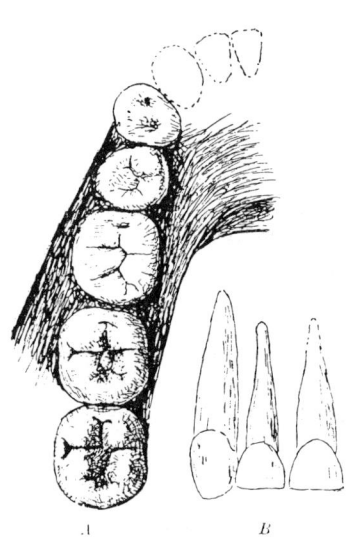

We have found two beautiful upper incisors and a perfect upper canine. These teeth are almost typically human, though a little larger than most human teeth. The canine has no deep infolding of the enamel on the lingual side as we have in the canines of *Plesianthropus*. It is also interesting to note that the canine and the second incisor have been in contact, as each has been abraded by the other.

The mandible was found in the same deposit but at a spot about 10 feet from the isolated upper teeth, and it cannot belong to the same individual though it is clearly of the same species.

Teeth of *Paranthropus crassidens* Broom. ¾ natural size. *A*. Left mandibular ramus. The 1st premolar is a little displaced in the specimen and has been restored to its natural position. The 3rd molar is drawn from the 3rd right molar reversed. *B*. Right upper incisors and canine.

The mandible is very massive. The horizontal ramus is preserved from the second premolar to the second molar; the first premolar is also preserved but a little displaced. Most of the inner side of the symphysis resembles more closely that of the Heidelberg jaw than any other specimen of man or ape-man I know.

But the teeth are relatively huge. The drawing is of the occlusal view with the first premolar in position, and the left third molar drawn from the right tooth reversed.

This new type of ape-man is not closely allied to either *Australopithecus* or *Plesianthropus*, but is allied to *Paranthropus*. When a skull is discovered it may prove to belong to a new genus; but provisionally we may call it *Paranthropus crassidens*.

As further evidence of the richness of our deposits, attention may be directed to the wonderful finds being made in the northern Transvaal at the Makapan caves. For about a couple of years the Bernard Price Institute has been working there, and Prof. R. A. Dart has recently announced the discovery of a remarkable ape-man occiput, and a few weeks later of a very fine mandible. These he has referred to a species of *Australopithecus*, and has called the animal *A. prometheus*. Though I am not convinced that he made fire, I am of opinion that the being belongs not only to a new species but also to a new genus. Prof. Dart has some other most important remains which he is describing. But I feel at liberty only to refer to what has been already announced.

8

Telanthropus capensis Broom and Robinson, 1949
Type specimen: SK 15

In this paper Robert Broom and John T. Robinson propose new genus and species names, *Telanthropus capensis*, based on a mandible from Swartkrans which they regard as distinctively different from the robust australopithecine majority of Swartkrans specimens. Most later workers have considered the type specimen, the mandible SK 15, as well as other non-australopithecine hominids from Swartkrans to be members of some species of the genus *Homo*. Other geologically ancient specimens of *Homo* from southern or eastern Africa are presumably conspecific with SK 15. If, however, such a hypothetical species of *Homo* is recognized it can never be called *"Homo capensis."* Ironically, this name would be invalid because of Broom's own actions in naming a species *Homo capensis* in 1917. The type specimen of *Homo capensis* Broom, 1917 is a skull from Boskop, South Africa, which is generally considered indistinguishable from living humans. Thus *Homo capensis* Broom, 1917 is a junior subjective synonym of *Homo sapiens*. Any later-named species *"capensis"*, when transferred to the genus *Homo*, becomes a junior homonym of Broom's 1917 name, and thus cannot be valid in *Homo*. Anyone who thought that SK 15 warranted placement in a non-*Homo* genus (although no one, to our knowledge, does so) would be free to use the name *Telanthropus capensis*. This is an example of the situation where a species name (in this case *"capensis* Broom and Robinson, 1949") is available, in nomenclatural terms, but is not valid within a particular genus (that is, *Homo*).

Note that this is the first taxon in this book to be proposed in a collaborative paper rather than by a single author. A teamwork approach has become much more common in paleoanthropology generally in recent decades, and later publications reflect the same trend in taxonomic publications (see selections 11–14).

A New Type of Fossil Man

Robert Broom and John T. Robinson

In the cave at Swartkrans which has now yielded the jaws and skulls of the huge ape-man *Paranthropus crassidens*, there was found by Mr. J. T. Robinson, on April 29, 1949, the lower jaw of what is fairly manifestly a new type of man. Though this was discovered in the same cave as the large ape-man, it is clearly of considerable later date. In the main bone breccia of the cave deposit there has been a pocket excavated and refilled by a darker type of matrix. The pocket was of very limited extent, being only about 4 feet by 3 feet and about 2 feet in thickness. The deposit was remarkably barren, there being no other bones in it except the human jaw and a few remains of very small mammals. We are thus at present unable to give the age of the deposit except to say that it must be considerably younger than the main deposit. If the main deposit is Upper Pliocene, not improbably the pocket may be Lower Pleistocene.

The jaw is smaller than many human jaws, though the third molar is larger than in any known man. On the left ramus the three molars are preserved in good condition though a little worn, and the last two molars are well preserved in the right ramus. No other teeth are preserved, though we have sockets of all of them.

The jaw has been a little broken during fossilization, and slightly crushed; but otherwise it is nearly perfect except for the loss of most of the left condyle and the whole of the right. A very small part of the lower symphyseal region is lost. The symphysis runs downwards and slightly backwards, making an angle with the base of the ramus of about 75°. The depth of the symphysis is about 33 mm. The horizontal ramus is remarkably shallow. At the first molar it is only 29 mm. The base of the ramus is nearly level, and the angle is rounded and scarcely at all below the general level.

The ascending ramus has apparently been fairly broad, but very shallow. Fortunately the cast of the side of the one condyle is

Reprinted by permission from *Nature*, Vol. 164, pp. 322–23. Copyright 1949 Macmillan Magazines, Ltd.

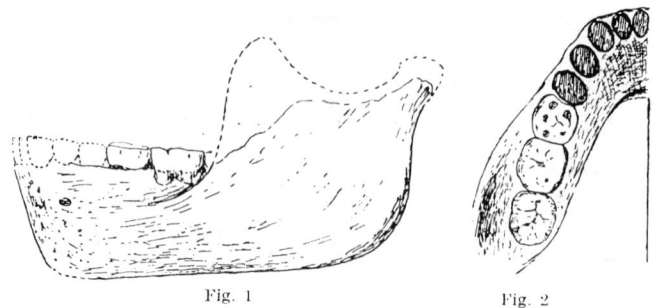

Fig. 1   Fig. 2

Figure 1.   Side view of lower jaw of *Telanthropus capensis*
B. and R. (half-size)

Figure 2.   Occlusal view of teeth of the left ramus of
Telanthropus capensis B. and R. (half-size)

Preserved, and the height of the back of the jaw is only about 55 mm. above the horizontal base of the jaw. Outside the last molar is a wide hollow as in the Heidelberg jaw, and the jaws of the ape-men. There is no simian shelf and the whole symphysis is not unlike that of Heidelberg man, but smaller. The mylohyoid groove runs up to the foramen as in typical human skulls. In *Paranthropus crassidens* the groove is, as in *Eoanthropus*, lower down.

The incisors and canines, so far as can be judged from the sockets, are human. The premolars have been a little larger than typically human premolars. The first molar is almost typically human in size and structure. It has five cusps and a trace of a sixth. The second molar is also nearly human. It is larger than in *Homo*, and has a small sixth cusp. The third molar is the largest of the three molars. It has five well-developed cusps and a small sixth.

The jaw in general structure comes nearest to that of Heidelberg man, but is smaller and has a lower horizontal ramus. The teeth differ markedly in the third molar, being the largest of the series.

In the large size of the molars there is some resemblance to the condition seen in *Plesianthropus* and *Paranthropus*; but in this human jaw the molars are much smaller. In *Plesianthropus transvaalensis* the three molars measure in the male about 43 mm.; in *Paranthropus robustus* they measure 45 mm. In *Paranthropus crassidens* the three measure in the male about 51 mm., while in this new human jaw they only measure 38.4 mm. In the South African native the molars measure about 35 mm.

The new type of man represented by this fossil jaw we propose to call *Telanthropus capensis*. We regard him as somewhat allied to Heidelberg man, and intermediate between one of the ape-men and true man.

It might be thought that as *Plesianthropus transvaalensis*, of which we now know about a dozen skulls and about a hundred and fifty teeth, shows considerable variation, this supposed human jaw might be an extreme variant of *Paranthropus crassidens*. In man there are no doubt great variations, and the difference in size between the jaw of a small Bushman woman and the Wadjak and Heidelberg jaws is nearly as great as between our supposed human jaw and the huge *Paranthropus crassidens* jaw. We now have three good lower jaws and a number of isolated teeth of *P. crassidens*, and there is not much variation in either size or structure. It may be held that all these large jaws are male jaws, and the small jaw that of a female; but not only the size of the teeth but also the structure seems to rule out such a view. The first molar in the type of *P. crassidens* is about 16 mm. by 14.6 mm. In the supposed human jaw it is only 12 mm. by 11.5 mm. Further, the structure of the two teeth differ considerably. The typically human mylohyoid groove in our supposed man, and the certainly not typically human groove in *Paranthropus*, seem to make it certain that the two jaws belong to different genera. If we are right in believing that our new jaw is in structure intermediate between *P. crassidens* and *Homo*, it is but natural that there should be numerous resemblances to both.

9

Meganthropus africanus Weinert, 1950
Type specimen: Garusi 1

Selection 9 is an excerpt from a longer paper by Hans Weinert in which he reviews a series of fossil primate and hominid discoveries of the 1930s and 1940s. The section reprinted here represents the first naming of a fossil hominid from the Laetoli area of Tanzania. A broken hominid canine had been collected at Laetoli by L. S. B. Leakey in 1935, but it was not recognized as a hominid until decades later (White, 1981). The specimen described and named by Weinert was found in 1939 by a German expedition led by Ludwig Kohl-Larsen. For a history of Kohl-Larsen's discovery and a detailed description of what has come to be known as the Garusi maxilla, see Protsch (1981). Weinert's publication marks the first description and naming of a hominid from Laetoli. Note that he only provisionally names it *Meganthropus africanus*. Such provisional actions were allowed before 1961, but are prohibited by the Code since then.

The species *Meganthropus palaeojavanicus* discussed by Weinert was created by G. H. R. von Koenigswald (see Campbell, selection 18). Its type specimen is a mandible from Sangiran, Java, found in 1941. The genus *Meganthropus* is commonly considered a synonym of *Homo*. In any case, the Garusi maxilla is not considered related to specimens from Java by any paleoanthropologist today; it is generally acknowledged as part of the same taxon as other Laetoli hominids recovered by M. D. Leakey in the 1970s. The Garusi maxilla was included (along with other Laetoli and Hadar specimens) in the hypodigm of *Australopithecus afarensis* by Johanson, White, and Coppens (see selection 14). Of course, when this type specimen is transferred to the genus *Australopithecus*, its potential name, "*Australopithecus africanus* (Weinert, 1950)", is a junior homonym of *Australopithecus africanus* Dart, 1925. This cannot be a valid name. Johanson, White, and Coppens could have designated "*Australopithecus afarensis*" as a replacement name for *Meganthropus africanus* Weinert, 1950, in the genus *Australopithecus*. If they had done so, the Garusi maxilla would automatically have become the type specimen of *Australopithecus afarensis* (following Article 72 (e) of the Code).

However, they did not explicitly take this approach, but instead designated a new holotype, the LH 4 mandible, for the species *afarensis*. However, in any genus other than *Australopithecus*, Weinert's name will retain its priority over that of Johanson, White, and Coppens. Therefore, if the Laetoli hominids, including the Garusi maxilla, are ever placed in some other genus, whether new or pre-existing, in which the name "*africanus*" is not a junior homonym, their proper species name will revert to "*africanus* (Weinert, 1950)." In fact, such a genus has already been named: *Praeanthropus* Senyürek, 1955 (see Senyürek, 1955). The type species of this genus is *Meganthropus africanus* Weinert, 1950, whose holotype is the Garusi maxilla. Therefore, if any future paleoanthropologist decides that all (or any portion which includes the Garusi maxilla) of *Australopithecus afarensis* belongs in a different genus, the name *Praeanthropus africanus* (Weinert, 1950) would be available for this taxon.

The specimens referred to by Weinert in this paper as *Africanthropus* are perhaps from Middle Pleistocene deposits, geologically much younger than the Pliocene Laetolil Beds which are the source of the Garusi maxilla. These specimens are today often considered similar to the Kabwe fossil (selection 4) in morphology and taxonomic position.

On the New Prehuman and Early Human Finds from Africa, Java, China, and France

Hans Weinert

The Maxilla Fragment with Premolars*
(*Meganthropus africanus*)

Forty km to the north of the site of *Africanthropus* I, Kohl-Larsen found on February 14, 1939, a broken right maxilla

* Since this is the first scientific announcement, I give the fragment provisionally the designation: "*Meganthropus africanus*."

From Hans Weinert, "Über die neuen Vor- und Frühmenschenfunde aus Afrika, Java, China und Frankreich," *Zeitschrift für Morphologie und Anthropologie*, Vol. 42, pp. 113–48 (1950). Translated by W. E. Meikle from "Das Oberkieferstück mit den Prämolaren," pp. 139–141 only, by permission of E. Schweizerbartsche Verlagsbuchhandlung, Stuttgart.

fragment with both premolars, and ten days later, about 6 km away, an isolated molar. Both these fragments must certainly be named together, for they both have a very similar appearance: light yellow, very heavily petrified, and extraordinarily dense; their specific gravity amounts to 2.7—close to that of the black fragments of *Africanthropus.*

Despite the external impression, however, it appears that we are not dealing with precisely the same forms. Although *Africanthropus* certainly has a quite considerable size, these specimens surely fall outside the bounds of other hominid finds: they are significantly larger. It was therefore good luck that I could compare them with the cast of the large *Meganthropus* from Java.

The first three illustrations for this section, Figures 12-1, 12-2, and 12-3, show the maxilla fragment with both premolars at natural size. Illustration 1 is an external view; on the right would have been the canine, of which a bit of the alveolus is still present. Illustration 2 shows this broken alveolus (filled with matrix in its upper part) in frontal view. Illustration 3 shows the fragment viewed from the inside. In illustrations 2 and 3 one can recognize the thickness of the maxilla at the palate and see, first of all, how low the curvature of the palate is. Moreover the right half of the nasal opening lacks any nasal spine and, exactly as in anthropoids, slopes into the upper alveolar border. From these morphological characteristics one would have to consider the fragment as the remains of an anthropoid if the teeth themselves did not prove their undoubted hominid nature.

The following two illustrations, Figures 12-4 and 13-1, show the specimen next to a modern *Homo sapiens.* One sees in the first illustration that the maxilla remains cannot be oriented at all precisely over the human mandible, since the fossil must project out in any view. In the frontal view the midline of the maxilla (marked with a vertical line in the figure) is not quite reached; one can thus reconstruct the piriform aperture. It is not especially broad, just as the entire maxilla of the fossil possesses no outstanding breadth. Unfortunately the corresponding left premolar is missing from the comparison figure with the *Homo sapiens.* For the left side of the skull one must therefore consult the mirror image for comparison.

It is further to be recognized that the base of the nasal opening lies no higher above the alveolar border than in the human skull. This is evidently at the expense of the enormous thickness of the palate of the African find.

The next four illustrations, Figures 13-2 and 13-3, and Figures 14-1 and 14-2, show the maxilla fragment in conjunction with the large *Meganthropus* II. These fit together. The African find is

Figure 12-1.   Tooth finds of
the Kohl-Larsen Expedition to
East Africa. Left, isolated upper
molar; right, maxilla fragment
with both premolars.

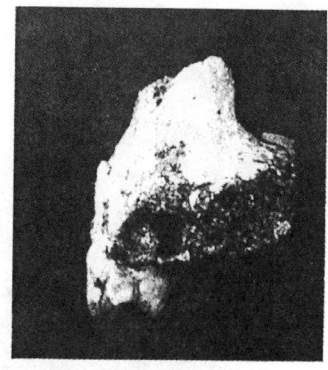

Figure 12-2.   The same
maxilla fragment viewed
from the front.

Figure 12-3.   The same
maxilla fragment viewed from
the inside.

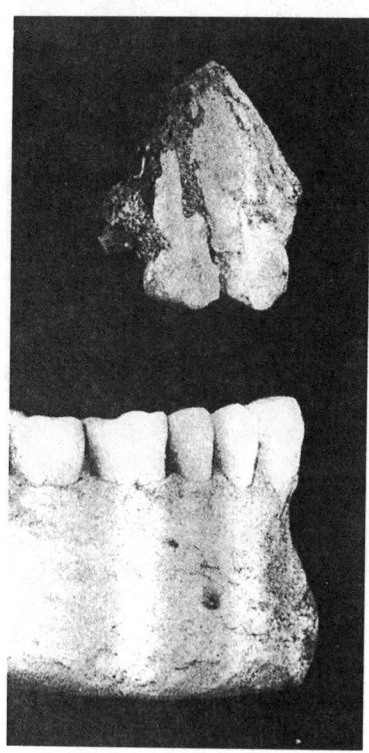

Figure 12-4.   The same
maxilla fragment in comparison
with the mandible of a strong
recent European.

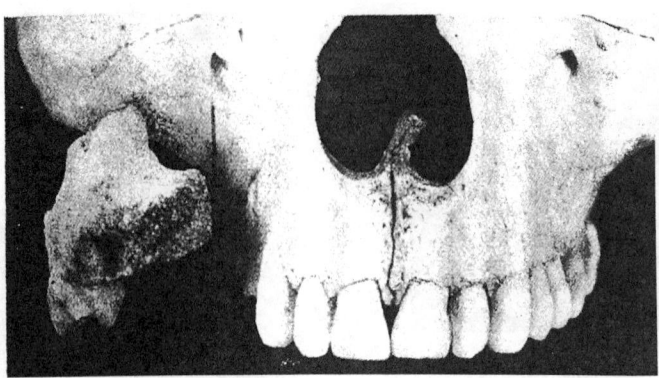

Figure 13-1.   The maxilla fragment from East Africa beside a large recent European.

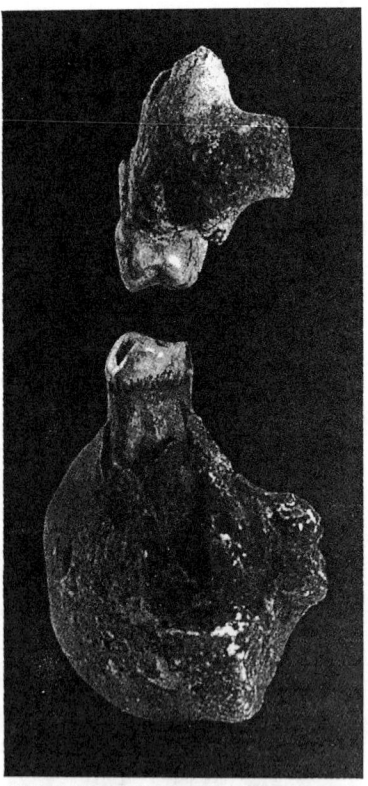

Figure 13-2.   The same maxilla fragment above the mandible of *Meganthropus* II, viewed from the front.

Figure 13-3.   The same fragments viewed from the back.

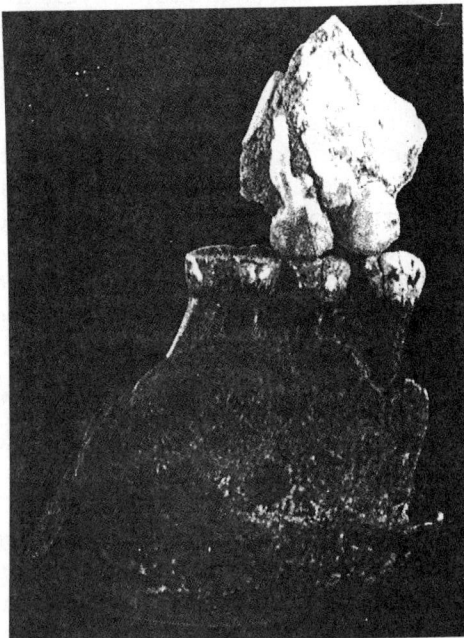

Figure 14-1.   The maxilla fragment from East Africa in correct position above the mandible of *Meganthropus* II, viewed from the outside.

Figure 14-2.   The same, viewed from the inside.

perhaps somewhat smaller than the Javanese one; however, the bone thickness is the same, and even the tooth size hardly differs between them. In any case, the African fossil fits with this *Meganthropus* better than with the other one which von Koenigswald has also discovered in Java.

Figure 14-1 shows the jaw fragments in external view. Although the teeth match each other with regard to the crown and root sizes, one would surely think that in *Meganthropus* the base of the nasal cavity would be higher above the occlusal plane.

Figure 14-2 produces the correct impression; however, here we have intentionally placed the upper jaw fragment too high because otherwise the teeth overlap. Besides, I want to show how high I imagine the nasal cavity of *Meganthropus* to be above the alveolar rim, from the size of the body of the mandible. Naturally we cannot safely assert this; even *Meganthropus* might have had a very low palate.

Figure 13-2 shows both fragments from the front and Figure 13-3 shows them from the back. The question is often raised whether the thickly swollen jaw of *Meganthropus* might not be a pathological formation; I don't believe this. Likewise Weidenreich, when he had the original in hand, specifically reported the same. In Figure 13-3, in posterior view, there surely is no impression of pathology. Here one recognizes more clearly than in the other figures that the tooth shapes of *Meganthropus* resemble those of an anthropoid more closely than do those of the African fossil: the crown height is small in proportion to the broad and robust roots in *Meganthropus*, and moreover the crowns are somewhat constricted superiorly.

With *Meganthropus* I we reach probably a still better fit to this fragment. However, no cast of this specimen is available. In any case, however, the African find is above the human range of variation in size and approaches that of *Meganthropus*, although it does not represent the same species in Africa. Broom has announced a completely similar mandible discovery from South Africa.

10

Zinjanthropus boisei Leakey, 1959
Type specimen: OH 5

In this paper Louis Leakey names a new genus and species, *Zinjanthropus boisei*, from Bed I of Olduvai Gorge, Tanzania. This is the first announcement of the holotype (Olduvai Hominid 5), which had been discovered less than a month before by Mary Leakey. The specimen was later described in great detail by Tobias (1967). At the time of naming this was the only known member of its species. However, in contrast to many other hominid species, the holotype is a largely complete, well-preserved, adult skull whose anatomy can easily be compared to that of others (contrast selections 2, 3, 7, 8, 9, 12, 13).

The affinities of Leakey's species with the robust australopithecines of South Africa have been universally recognized, despite his original attempts to distinguish it from all the other australopithecines. No doubt the presence of Oldowan artifacts was significant in influencing Leakey's view on the importance of this find. To see how the discoveries starting in 1960 of fossils later named *Homo habilis* changed Leakey's views on the significance of OH 5 and of *Zinjanthropus boisei* (Tobias, 1991), compare this article with that presented in the following selection (11).

The discovery of Olduvai Hominid 5 had more significant consequences for paleoanthropology than any other fossil find in this century except for Taung. Following the recovery of this first largely complete australopithecine skull from East Africa there was a marked increase of interest and funding for paleoanthropological research in eastern Africa. This growth accelerated through the 1960s. Increased funding for the Leakeys' work at Olduvai led directly to further discoveries by them (again, see selection 11). Success at Olduvai helped make possible expeditions by others and thus contributed, directly and indirectly, to such important fieldwork as that at the Omo (which led in turn to Koobi Fora), at Hadar, and at Laetoli.

Another way to appreciate the historical significance of the OH 5 discovery is to consider the number of hominid type

85

specimens later discovered by expeditions associated somehow with
the Leakey family. OH 7 (see selection 11) was found by Jonathan
Leakey at Olduvai during subsequent work by his parents, Louis and
Mary Leakey. Omo 18-(1967)-18 (selection 12) was found in the Omo
by the French portion of a joint American-Ethiopian-French-Kenyan
expedition. Richard Leakey led the Kenyan contingent that year. KNM-
ER 992 (selection 13) was found in the Ileret area east of Lake
Turkana during fieldwork led by Richard Leakey. LH 4 (selection 14)
was found at Laetoli during Mary Leakey's research there. KNM-ER
1470 (selection 15) was also found during Richard Leakey's work east
of Lake Turkana, at Koobi Fora.

A New Fossil Skull from Olduvai

L. S. B. Leakey

On July 17, at Olduvai Gorge in Tanganyika Territory, at Site *FLK*,
my wife found a fossil hominid skull, at a depth of approximately
22 ft. below the upper limit of Bed I. The skull was in the process
of being eroded out on the slopes, and it was only because this
erosion had already exposed part of the specimen that the discovery
was possible. Excavations were begun on the site the following day
and continued until August 6. As a result, an almost complete skull
of a hominid was discovered. This skull was found to be associated
with a well-defined living floor of the Oldowan, pre-Chelles-Acheul,
culture.

Upon the living floor, in addition to Oldowan tools and waste
flakes, there were fossilized broken and splintered bones of the
animals that formed part of the diet of the makers of this most
primitive stone-age culture. It has not yet been possible to study
the fauna found on this living floor; but it can be said that it includes
birds, amphibians, reptiles such as snakes and lizards, many
rodents and also immature examples of two genera of extinct pigs,
as well as antelope bones, jaws and teeth.

It is of special importance to note that whereas the bones of the larger animals have all been broken and scattered, the hominid skull was found as a single unit within the space of approximately one square foot by about six inches deep. Even fragile bones like the nasals are preserved. The expansion and contraction of the bentonitic clay, upon which the skull rested and in which it was partly embedded, had resulted, over the years, in its breaking up into small fragments which have had to be pieced together. The bones, however, are not in any way warped or distorted. A large number of fragments still remain to be pieced together.

This very great difference between the condition of the hominid skull and that of the animal bones on the same living floor (all of which had been deliberately broken up) seems to indicate clearly that this skull represents one of the hominids who occupied the living site; who made and used the tools and who ate the animals. There is no reason whatever, in this case, to believe that the skull represents the victim of a cannibalistic feast by some hypothetical more advanced type of man. Had we found only fragments of skull, or fragments of jaw, we should not have taken such a positive view of this.

It therefore seems that we have, in this skull, an actual representative of the type of 'man' who made the Oldowan pre-Chelles-Acheul culture.

This skull has a great many resemblances to the known members of the sub-family of Australopithecinae. Some scientists recognize only one genus, namely, *Australopithecus*, and treat Broom's *Paranthropus* as a synonym; others consider that the demonstrable differences are of such a nature that both genera are valid. Personally, having recently re-examined all the material of the two genera, in Johannesburg and Pretoria, I accept both as valid.

The Olduvai skull is patently a member of the sub-family Australopithecinae, and in certain respects it recalls the genus *Paranthropus*. In particular, this is the case in respect of the presence of the sagittal crest, the great reduction in the size of the canines and the incisors, the relatively straight line of these teeth at the front of the palate, the position of the nasal spines and the flatness of the forehead. In certain other characters, the new skull resembles more closely the genus *Australopithecus*, for example in respect of the high cranial vault, the deeper palate and the reduction of the upper third molars to a size smaller than the second, all of which are features to be found in *Australopithecus* but not in *Paranthropus*.

The very close examination and direct comparisons which I have personally made in South Africa have convinced me that, on the basis of our present state of knowledge, the new skull from Olduvai, while clearly a member of the Australopithecinae, differs from both

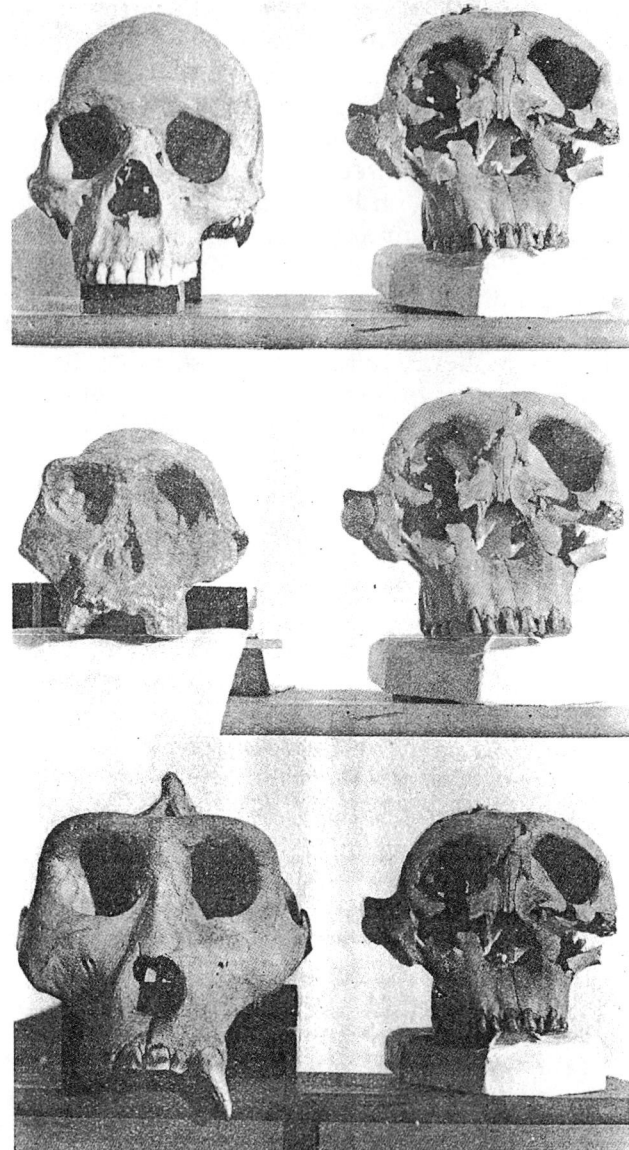

Figure 1. *Above*: The new skull compared with the skull of an Australian aboriginal. Note the very long face, the architecture of the malar region, the unusual nasal bones, the torus above the mastoid, the sagittal and nuchal crests.

Middle: The new skull compared with a cast of the most complete adult of *Australopithecus*. Note the difference in the size and shape of the face, the shape of the tympanic plate, the low position of the inion, the huge mastoid, as well as the difference in the shape of the malar region and the supra-orbital area.

Below: The new skull seen next to that of a gorilla.

Figure 2.   The palate of the new skull compared with that of an East African native.

Australopithecus and *Paranthropus* much more than these two genera differ from each other.

I am not in favour of creating too many new generic names among the Hominidae; but I believe that it is desirable to place the new find in a separate and distinct genus. I therefore propose to name the new skull *Zinjanthropus boisei*. This generic name derives from the word 'Zinj', which is the ancient name for East Africa as a whole, while the specific name is in honour of Mr. Charles Boise, whose constant encouragement and financial help ever since 1948 have made this and other important discoveries possible. I would also like to acknowledge the generous help received, from time to time, from the Wenner-Gren Foundation and the Wilkie Trust.

The following is the preliminary diagnosis of the new genus and the new species:

Zinjanthropus gen. nov.:

Genotype: a young male with third molars not yet in wear and sutures relatively open, from *FLK* I, Olduvai.

A new genus of the Hominidae, sub-family Australopithecinae, which exhibits the following major differences from the genera

Australopithecus and *Paranthropus*:

(*a*) in males a nuchal crest is developed as a continuous ridge across the occipital bone;

(*b*) the inion, despite the great evidence of muscularity, is set lower (when the skull is in the Frankfurt plane) than in the other two genera;

(*c*) the posterior wall of the occipital bone rises more steeply to form, with the parietals, a very high-vaulted posterior region of the skull;

(*d*) the foramen magnum is less elongate and has a more horizontal position than in *Australopithecus* (in the crushed skulls of *Paranthropus* it is not possible to be quite sure of the plane of the foramen magnum);

(*e*) the presence of a very massive horizontal ridge or torus above the mastoids. This is much more marked than the normal type of supra-mastoid crest;

(*f*) the mastoids are more similar to those seen in present-day man, both in size and shape;

(*g*) the presence of a strong wide shelf above the external auditory meatus, posterior to the jugal element of the temporal bone;

(*h*) the shape and form of the tympanic plate, whether seen in *norma lateralis* or in *norma basalis*. In this character the new skull has similarities with the Far Eastern genus *Pithecanthropus*;

(*i*) the very great pneumatosis of the whole of the mastoid region of the temporal bones, which even invades the squamosal elements;

(*j*) the massiveness of the jugal element of the temporal bone relative to the total size of the temporal bone;

(*k*) the way in which the parietals rise almost vertically behind the squamous elements of the temporal before bending over to become a dome;

(*l*) the relative thinness of the parietals in comparison with the occipitals and the temporals;

(*m*) the very prominent and keeled anterior margin of the crests on the frontal bone for the anterior segment of the temporal muscles in the region of the post-orbital constriction (even the most muscular male *Paranthropus* exhibits nothing comparable);

(*n*) the very unusual position of the nasion, which is on the most anterior part of the skull, instead of being behind and below the glabella region;

[Sic: No item lettered "o" appeared in the original article.]

(*p*) the very great absolute and also relative width of the inter-orbital area, with which may be associated the shape of the nasal bones, which are much wider at the top than at their inferior margin;

(*q*) the whole shape and position of the external orbital angle

elements of the frontal bone;

(r) the very deep palate which is even more markedly like that of *Homo* than in *Australopithecus*, and is quite unlike the form seen in *Paranthropus*, except in respect of the more or less straight canine-incisor line which has already been commented on, as a character recalling *Paranthropus*;

(s) the conformation of the malar-maxillary area of the cheek. In all known members of the genera *Australopithecus* and *Paranthropus* there is a buttress of bone which runs down from the malar towards the alveolar margin of the maxilla in about the region of the fourth premolar; in *Zinjanthropus* this buttress is wholly absent and the form of architecture of this region is that which is found in *Homo*;

(t) the very great area of muscle attachment on the inferior margin of the malars;

(u) the relatively greater reduction of the canines in comparison with the molar-premolar series than is seen even in *Paranthropus*; where it is a marked character.

Zinjanthropus boisei sp. nov.

A species of *Zinjanthropus* in which the males are far more massive than the most massive male *Paranthropus*. The face is also excessively long. Males have a sagittal crest, at least posteriorly. Upper third molars smaller than the second.

The above is only a preliminary diagnosis of the genus *Zinjanthropus* species *boisei*. It is recognized that, if and when further material is found, the diagnosis will need both enlarging and possibly modifying.

The whole question of generic value is one which is relative. There are some who maintain that *Australopithecus* and *Paranthropus* are not generically distinct, and who will wish to treat *Zinjanthropus* as a third, but less specialized, species of a single genus; but the differences seem to be too great for this.

I must now turn to the absolute and relative geological age of the new skull. As stated earlier, *Zinjanthropus* comes from Olduvai Gorge, about 22 ft. below the upper limit of Bed I. It was found in association with tools of the Oldowan culture, on a living floor and with associated fauna.

In the past it has been customary to regard Olduvai Bed I as a part of the Middle Pleistocene, not differentiating it from Bed II. During the last few years, however, detailed excavations at sites *BK* II, *SHK* II and *HWK* II have shown that there is a constant and well-marked break between the top of Bed I and the base of Bed II. It is incidentally on this clearly defined land surface that Chellean Stage I living sites are found.

There has also been found a great deal of new faunal evidence, and it is now clear that the fauna of Olduvai Bed I is the same as that of Omo, and that both are generally of the same age as that of Taungs. In other words, it is now necessary to regard Olduvai Bed I as representing the upper half of the Villafranchian and not the lower part of the Middle Pleistocene. So far as relative dating is concerned, it now seems clear that in the Far East the Djetis beds belong to the Middle, rather than to the Lower, Pleistocene, so that the new Olduvai skull would be older than the oldest *Pithecanthropus.*

In South Africa, the deposits at Taungs and Sterkfontein are now regarded as belonging to the upper part of the Lower Pleistocene; they must therefore be regarded as generally contemporary with Olduvai Bed I. The Makapan beds are a little younger, in all probability, while Swartkrans is of Middle Pleistocene age, as are the upper beds at Sterkfontein which are now yielding stone tools.

With the Taungs child, therefore, and the *Australopithecus* fossils from the lower beds at Sterkfontein, the new find represents one of the earliest Hominidae, with the Olduvai skull as the oldest yet discovered maker of stone tools.

The following approximate measurements will indicate the size of the new specimen.

Length from inion to glabella	about 174 mm.
Greatest breadth at supra-mastoid torus	about 138 mm.
Greatest breadth of brain case on squamosal element of the temporal bones	about 118 mm.
Height (in Frankfurt plane) from basion to a point vertically above it in the sagittal plane	about 98 mm.
External orbital angle width	about 122 mm.
Inter-orbital width	about 32.5 mm.
Post-orbital width	about 88 mm.
Palate-length from front of incisors to a line joining back of third molars	about 84 mm.
Palate-width at second molars	about 82 mm.
Palate-width at third premolars	about 62 mm.
Length of molar-premolar series	about 72 mm.

Teeth measurements:
$M3$: 21 x 16 mm.; $M2$: 21 x 17 mm.; $M1$: 18 x 15.5 mm.; $PM4$: 18 x 12 mm.; $PM3$: 17 x 11.5 mm.; C: 9.5 x 9 mm.; $I2$: 7 x 7 mm.; $I1$ (both damaged but about 10 x 8 mm.).

11

Homo habilis Leakey, Tobias and Napier, 1964
Type specimen: OH 7

Following OH 5 (selection 10), the continuing recovery of specimens at Olduvai Gorge through the early 1960s led to the recognition and naming of a second, non-australopithecine, hominid species by L. S. B. Leakey, P. V. Tobias and J. R. Napier. The naming of *Homo habilis* led to a protracted and multisided argument about the boundaries of the genus *Homo* (Tobias, 1991). Eventually *H. habilis* became generally accepted, to the point that the sequence of species *H. habilis—H. erectus—H. sapiens* within the *Homo* lineage is almost a textbook cliché. This came about less through published analysis, comparison, or argument than by the cumulative impact of the non-australopithecine "early *Homo*" specimens recovered since about 1970 at Koobi Fora and elsewhere (Wood, 1991). Now, ironically coinciding with publication of the detailed monographic treatment of the original Olduvai hypodigm of *H. habilis* (Tobias, 1991), new arguments have appeared. No longer is the debate over whether creation of this species was justified thirty years ago; it clearly was. Now the question is not if one pre- or non-*Homo erectus* species of "early *Homo*" should be recognized, but how many more there might be besides *Homo habilis* (Wood, 1991).

Note the careful, formal structure of the taxonomic and nomenclatural sections of this paper, in contrast to most of the early works reprinted in this book. In terms of contents and presentation (fossil diagnosis; description; hypodigm; discussion) this publication fulfills all the requirements and expectations that one might have for a very short, succinct presentation of a new taxon and name in paleoanthropology. (The structure of the Johanson, White and Coppens paper (selection 14) is also exemplary.)

When published, *Homo habilis* represented both a newly recognized taxon and a new name. Part of its original hypodigm had been described in earlier papers in *Nature*, and part was first described here.

A New Species of the Genus *Homo* from Olduvai Gorge

L. S. B. Leakey, P. V. Tobias and J. R. Napier

The recent discoveries of fossil hominid remains at Olduvai Gorge have strengthened the conclusions—which each of us had reached independently through our respective investigations—that the fossil hominid remains found in 1960 at site *F.L.K.N.N.* I, Olduvai, did not represent a creature belonging to the sub-family Australopithecinae.[1]

We were preparing to publish the evidence for this conclusion and to give a scientific name to this new species of the genus *Homo*, when the new discoveries which are described by L. S. B. and M. D. Leakey in the preceding article [in this same issue of *Nature*], were made.

An examination of these finds has enabled us to broaden the basis of our diagnosis of the proposed new species and has fully confirmed the presence of the genus *Homo* in the lower part of the Olduvai geological sequence, earlier than, contemporary with, as well as later than, the *Zinjanthropus* skull, which is certainly an australopithecine.

For the purpose of our description here, we have accepted the diagnosis of the family Hominidae, as it was proposed by Sir Wilfrid Le Gros Clark in his book *The Fossil Evidence for Human Evolution* (110; 1955). Within this family we accept the genus *Australopithecus* with, for the moment, three sub-genera (*Australopithecus*, *Paranthropus* and *Zinjanthropus*) and the genus *Homo*. We regard *Pithecanthropus* and possibly also *Atlanthropus* (if it is indeed distinct) as species of the genus *Homo*, although one of us (L. S. B. L.) would be prepared to accept sub-generic rank.

It has long been recognized that as more and more discoveries were made, it would become necessary to revise the diagnosis of the genus *Homo*. In particular, it has become clear that it is impossible to rely on only one or two characters, such as the cranial capacity or an erect posture, as the necessary criteria for membership of the genus. Instead, the total picture presented by the material available for investigation must be taken into account.

We have come to the conclusion that, apart from *Australopithecus* (*Zinjanthropus*), the specimens we are dealing with from Bed I and the lower part of Bed II at Olduvai represent a single species of the genus *Homo* and not an australopithecine. The species is, moreover, clearly distinct from the previously recognized species of the genus. But if we are to include the new material in the genus *Homo* (rather than set up a distinct genus for it, which we believe to be unwise), it becomes necessary to revise the diagnosis of this genus. Until now, the definition of *Homo* has usually centered about a "cerebral Rubicon" variably set at 700 c.c. (Weidenreich), 750 c.c. (Keith), and 800 c.c. (Vallois). The proposed new definition follows:

Family   HOMINIDAE (as defined by Le Gros Clark, 1955)

Genus   *Homo* Linnæus

Revised Diagnosis of the Genus Homo. A genus of the Hominidae with the following characters: the structure of the pelvic girdle and of the hind-limb skeleton is adapted to habitual erect posture and bipedal gait; the fore-limb is shorter than the hind-limb; the pollex is well developed and fully opposable and the hand is capable not only of a power grip but of, at the least, a simple and usually well developed precision grip[2]; the cranial capacity is very variable but is, on the average, larger than the range of capacities of members of the genus *Australopithecus*, although the lower part of the range of capacities in the genus *Homo* overlaps with the upper part of the range in *Australopithecus*; the capacity is (on the average) large relative to body-size and ranges from about 600 c.c. in earlier forms to more than 1,600 c.c.; the muscular ridges on the cranium range from very strongly marked to virtually imperceptible, but the temporal crests or lines never reach the midline; the frontal region of the cranium is without undue post-orbital constriction (such as is common in members of the genus *Australopithecus*); the supra-orbital region of the frontal bone is very variable, ranging from a massive and very salient supra-orbital torus to a complete lack of any supra-orbital projection and a smooth brow region; the facial skeleton varies from moderately prognathous to orthognathous, but it is not concave (or dished) as is common in members of the

Australopithecinae; the anterior symphyseal contour varies from a marked retreat to a forward slope, while the bony chin may be entirely lacking, or may vary from a slight to a very strongly developed mental trigone; the dental arcade is evenly rounded with no diastema in most members of the genus; the first lower premolar is clearly bicuspid with a variably developed lingual cusp; the molar teeth are variable in size, but in general are small relative to the size of these teeth in the genus *Australopithecus*; the size of the last upper molar is highly variable, but it is generally smaller than the second upper molar and commonly also smaller than the first upper molar; the lower third molar is sometimes appreciably larger than the second; in relation to the position seen in the Hominoidea as a whole, the canines are small, with little or no overlapping after the initial stages of wear, but when compared with those of members of the genus *Australopithecus*, the incisors and canines are not very small relative to the molars and premolars; the teeth in general, and particularly the molars and premolars, are not enlarged bucco-lingually as they are in the genus *Australopithecus*; the first deciduous lower molar shows a variable degree of molarization.

Genus   *Homo*   Linnæus
Species   *habilis*   sp. nov.

(NOTE: The specific name is taken from the Latin, meaning "able, handy, mentally skillful, vigorous." We are indebted to Prof. Raymond Dart for the suggestion that *habilis* would be a suitable name for the new species.)

A species of the genus *Homo* characterized by the following features:

A mean cranial capacity greater than that of members of the genus *Australopithecus*, but smaller than that of *Homo erectus*; muscular ridges on the cranium ranging from slight to strongly marked; chin region retreating, with slight or no development of the mental trigone; maxillæ and mandibles smaller than those of *Australopithecus* and within the range for *Homo erectus* and *Homo sapiens*; dentition characterized by incisors which are relatively large in comparison with those of both *Australopithecus* and *Homo erectus*; canines which are proportionately large relative to the premolars; premolars which are narrower (in bucco-lingual breadth) than those of *Australopithecus*, but which fall within the range for *Homo erectus*; molars in which the absolute dimensions range between the lower part of the range in *Australopithecus* and the upper part of the range in *Homo erectus*; a marked tendency towards bucco-lingual narrowing and mesiodistal elongation of all the teeth, which is especially evident in the lower premolars (where it expresses itself as a marked elongation of the talonid) and in the lower molars (where it is accompanied by a rearrangement of the distal cusps); the sagittal curvature of the parietal bone varies from slight (within the hominine range) to moderate (within the australopithecine range); the external sagittal curvature of the occipital bone is slighter than in *Australopithecus* or in *Homo erectus*, and lies within the range of *Homo sapiens*; in curvature as well as in some other morphological traits, the clavicle resembles, but is not identical to, that of *Homo sapiens sapiens*; the hand bones differ from those of *Homo sapiens sapiens* in robustness, in the dorsal curvature of the shafts of the phalanges, in the distal attachment of *flexor digitorum superficialis*, in the strength of fibro-tendinous markings, in the orientation of trapezium in the carpus, in the form of the scaphoid and in the marked depth of the carpal tunnel; however, the hand bones resemble those of *Homo sapiens sapiens* in the presence of broad, stout, terminal phalanges on fingers and thumb, in the form of the distal articular surface of the capitate and the ellipsoidal form of the metacarpo-phalangeal joint surfaces; in many of their characters the foot bones lie within the range of variation of *Homo sapiens sapiens*; the hallux is stout, adducted and plantigrade; there are well-marked longitudinal and transverse arches; on the other hand, the 3rd metatarsal is relatively more robust than it is in modern man, and there is no marked difference in the radii of curvature of the medial and lateral profiles of the trochlea of the talus.

Geological Horizon. Upper Villafranchian and Lower Middle Pleistocene.

Type. The mandible with dentition and the associated upper molar, parietals and hand bones, of a single juvenile individual from site *F.L.K.N.N.* I, Olduvai, Bed I.

This is catalogued as Olduvai Hominid 7.

Paratypes. (a) An incomplete cranium, comprising fragments of the frontal, parts of both parietals, the greater part of the occipital, and parts of both temporals, together with an associated mandible with canines, premolars and molars complete on either side but with the crowns of the incisors damaged, parts of both maxillæ, having all the cheek teeth except the upper left fourth premolar. The condition of the teeth suggests an adolescent. This specimen, from site *M.N.K.* II, Olduvai, Bed II, is catalogued as Olduvai Hominid 13.

(b) The associated hand bones, foot bones and probably the clavicle, of an adult individual from site *F.L.K.N.N.* I, Olduvai, Bed I. This is catalogued as Olduvai Hominid 8.

(c) A lower premolar, an upper molar and cranial fragments from site *F.L.K.* I, Olduvai, Bed I (the site that yielded also the *Australopithecus (Zinjanthropus)* skull). This is catalogued as Olduvai Hominid 6. (It is possible that the tibia and fibula found at this site belong with *Homo habilis* rather than with *Australopithecus (Zinjanthropus)*. These limb bones have been reported on by Dr. P. R. Davis (*Nature*, March 7, 1964, p. 967).

(d) A mandibular fragment with a molar in position and associated with a few fragments of other teeth from site *M.K.* I., Olduvai, Bed I. This specimen is catalogued as Olduvai Hominid 4.

Description of the Type. Preliminary descriptions of the specimens which have now been designated the type of *Homo habilis*, for example, the parts of the juvenile found at site *F.L.K.N.N.* I in 1960, have already been published in *Nature* by one of us (**189**, 649; **191**, 417; 1961). A further detailed description and report on the parietals, the mandible and the teeth are in active preparation by one of us (P. V. T.), while his report on the cranial capacity (preceding article) as well as a preliminary note on the hand by another of us (*Nature*, **196**, 409; 1962) have been published. We do not propose, therefore, to give a more detailed description of the type here.

Description of the Paratypes. A preliminary note on the clavicle and on the foot of the adult, which represents paratype (b), was published in *Nature* (**188**, 1050; 1960), and a further report on the foot by Dr. M. H. Day and Dr. J. R. Napier was published in *Nature* of March 7, 1964, p. 969.

The following additional preliminary notes on the other paratypes have been prepared by one of us (P. V. T.).

Description of Paratypes

(a) Olduvai Hominid 13 from M.N.K. *II.* An adolescent represented by a nearly complete mandible with complete, fully-erupted lower dentition, a right maxillary fragment including palate and all teeth from P^3 to M^3, the latter in process of erupting; the corresponding left maxillary fragment with M^1 to M^3, the latter likewise erupting, the isolated left P^3; parts of the vault of a small, adult cranium, comprising much of the occipital, including part of the posterior margin of *foramen magnum*, parts of both parietals, right and left temporosphenoid fragments, each including the mandibular fossa and foramen ovale. The distal half of a humeral shaft (excluding the distal extremity) may also belong to Olduvai Hominid 13. The *corpus mandibulae* is very small, both the height and thickness at M_1 falling below the australopithecine range and within the hominine range. All the teeth are small compared with

those of Australopithecinae, most of the dimensions falling at or below the lower extreme of the australopithecine ranges. On the other hand, practically all the dental dimensions can be accommodated within the range of fossil Homininae. The Olduvai Hominid 13 teeth show the characteristic mesiodistal elongation and labiolingual narrowing, in some teeth the *L/B* index exceeding even those of the type Olduvai Hominid 7, and paratype Olduvai Hominid 6. The occipital bone has a relatively slight sagittal curvature, the Occipital Sagittal Index being outside the range for australopithecines and for *Homo erectus pekinensis* and within the range for *Homo sapiens*. On the other hand, the parietal sagittal curvature is more marked than in all but one australopithecine and in all the Pekin fossils, the index falling at the top of the range of population means for modern man. Both parietal and occipital bones are very small in size, being exceeded in some dimensions by one or two australopithecine crania and falling short in all dimensions of the range for *Homo erectus pekinensis*. The form of the parietal—anteroposteriorly elongated and bilaterally narrow, with a fairly abrupt lateral descent in the plane of the parietal boss—reproduces closely these features in the somewhat larger parietal of the type specimen (Olduvai Hominid 7 from *F.L.K.N.N.* I).

(b) *Olduvai Hominid 6 from* F.L.K I. An unworn lower left premolar, identified as P_3, an unworn, practically complete crown and partly developed roots of an upper molar, either M^1 or M^2, as well as a number of fragments of cranial vault. These remains were found at the *Zinjanthropus* site and level, some *in situ* and some on the surface. Both teeth are small for an australopithecine, especially in buccolingual breadth, but large for *Homo erectus*. The marked tendency to elongation and narrowing imparts to both teeth an *L/B* index outside the range for all known australopithecine homologues and even beyond the range for *Homo erectus pekinensis*. The elongating-narrowing tendency is more marked in this molar than in the upper molar belonging to the type specimen (Olduvai Hominid 7) from *F.L.K.N.N.* I.

(c) *Olduvai Hominid 8 from* F.L.K.N.N. I. Remains of an adult individual found on the same horizon as the type specimen, and represented by two complete proximal phalanges, a fragment of a rather heavily worn tooth (premolar or molar), and a set of foot-bones possessing most of the specializations associated with the plantigrade propulsive feet of modern man. Probably the clavicle found at this site belongs to this adult rather than to the juvenile type-specimen; it is characterized by clear overall similarities to the clavicle of *Homo sapiens sapiens*.

(d) *Olduvai Hominid 4 from* M.K. I. A fragment of the posterior part of the left *corpus mandibulae*, containing a well-preserved, fully erupted molar, either M_2 or M_3. The width of the mandible is 19.2 mm, level with the mesial half of the molar, but the maximum width must have been somewhat greater. The molar is 15.1 mm in mesiodistal length and 13.0 mm in buccolingual breadth; it is thus a small and narrow tooth by australopithecine standards, but large in comparison with *Homo erectus* molars. There are several other isolated dental fragments, including a moderately worn molar fragment. These are stratigraphically the oldest hominid remains yet discovered at Olduvai.

Referred Material

Olduvai Hominid 14 from M.N.K. *II.* (1) A juvenile represented by a fragment of the right parietal with clear, unfused sutural margins; two smaller vault fragments with sutural margins; a left and a right temporal fragment, each including the mandibular fossa.

(2) A fragmentary skull with parts of the upper and lower dentition of a young adult from site *F.L.K.* II, Maiko Gully, Olduvai, Bed II, is also provisionally referred to *Homo habilis*. This specimen is catalogued as Olduvai Hominid 16. It is represented by the complete upper right dentition, as well as some of the left maxillary teeth, together with some of the mandibular teeth. The skull fragments include parts of the frontal, with both the external orbital angles preserved, as well as the supra-orbital region, except for the glabella; parts of both parietals and the occipital are also represented.

Implications for Hominid Phylogeny

In preparing our diagnosis of *Homo habilis*, we have not overlooked the fact that there are several other African (and perhaps Asian) fossil hominids whose status may now require re-examination in the light of the new discoveries and of the setting up of this new species. The specimens originally described by Broom and Robinson as *Telanthropus capensis* and which were later transferred by Robinson to *Homo erectus* may well prove, on closer comparative investigation, to belong to *Homo habilis*. The Kanam mandibular fragment, discovered by the expedition in 1932 by one of us (L. S. B. L.), and which has been shown to possess archaic features (Tobias, *Nature*, **185**, 946; 1960), may well justify further investigation along these lines. The Lake Chad craniofacial fragment, provisionally described by M. Yves Coppens in 1962, as an australopithecine, is not, we are convinced, a member of this sub-family. We understand that the discoverer himself, following his investigation of the australopithecine originals from South Africa and Tanganyika, now shares our view in this respect. We believe that it is very probably a northern representative of *Homo habilis*.

Outside Africa, the possibility will have to be considered that the teeth and cranial fragments found at Ubeidiyah on the Jordan River in Israel may also belong to *Homo habilis* rather than to *Australopithecus.*

Cultural Association

When the skull of *Australopithecus (Zinjanthropus) boisei* was found on a living floor at *F.L.K.* I, no remains of any other type of hominid were known from the early part of the Olduvai sequence. It seemed reasonable, therefore, to assume that this skull represented the makers of the Oldowan culture. The subsequent discovery of remains of *Homo habilis* in association with the Oldowan culture at three other sites has considerably altered the position. While it is possible that *Zinjanthropus* and *Homo habilis* both made stone tools, it is probable that the latter was the more advanced tool maker and that the *Zinjanthropus* skull represents an intruder (or a victim) on a *Homo habilis* living site.

The recent discovery of a rough circle of loosely piled stones on the living floor at site *D.K.* I, in the lower part of Bed I, is noteworthy. This site is geologically contemporary with *M.K.* I, less than one mile distant, where remains of *Homo habilis* have been found. It seems that the early hominids of this period were capable of making rough shelters or windbreaks and it is likely that *Homo habilis* may have been responsible.

Relationship to *Australopithecus (Zinjanthropus)*

The fossil human remains representing the new species *Homo habilis* have been found in Bed I and in the lower and middle part of Bed II. Two of the sites, *M.K.* I and *F.L.K.N.N.* I, are geologically older than that which yielded the skull of the australopithecine *Zinjanthropus.* One site, *F.L.K.* I, has yielded both *Australopithecus (Zinjanthropus)* and remains of *Homo habilis*, while two sites are later, namely *M.N.K.* II and *F.L.K.* II Maiko gully. The new mandible of *Australopithecus (Zinjanthropus)* type from Lake Natron, reported in the preceding article by Dr. and Mrs. Leakey, was associated with a fauna of Bed II affinities.

It thus seems clear that two different branches of the Hominidae were evolving side by side in the Olduvai region during the Upper Villafranchian and the lower part of the Middle Pleistocene.

Notes

[1] See also *Nature* of March 7 [1964], pp. 967, 969, and preceding articles in this issue.
[2] For the definition of "power grip" and "precision grip," see Napier, J. R., *J. Bone and Joint Surg.*, **38**, B, 902 (1956).

12

Paraustralopithecus aethiopicus Arambourg and Coppens, 1968
Type specimen: Omo 18-(1967)-18

In this paper Camille Arambourg and Yves Coppens create new genus and species names based on one specimen, a toothless mandible from the Shungura Formation, near the Omo River in southwestern Ethiopia. Actually, they first proposed their new names the year before (Arambourg and Coppens, 1967), but because that paper only proposed the names "provisionally," the correct citation for nomenclatural purposes (following Article 15 of the Code) is this paper from the *South African Journal of Science*. In the years following its publication this species name was usually considered a synonym of *Australopithecus boisei* or *Australopithecus africanus*, while the genus *Paraustralopithecus* has never been generally adopted. Following the discovery of KNM-WT 17000 ("the Black Skull") in 1985 west of Lake Turkana, several paleoanthropologists have revived the species name *aethiopicus* to express their opinion that WT 17000 represents the same distinct species as Arambourg and Coppens's mandible. Others regard WT 17000 as an early member of the *Australopithecus boisei* lineage and not as a separate species; they continue to regard *Australopithecus aethiopicus* as a synonym of *Australopithecus boisei*. See papers in Grine (1988a) for discussion of this unsettled controversy.

Discovery of a New Australopithecine in the Omo Beds (Ethiopia)

C. Arambourg and Y. Coppens

The exploration in 1933 by one of us (C.A.) of the Pleistocene beds of the Omo Valley (Ethiopia) revealed the extraordinary paleontological richness of these beds, and permitted, at the same time, the determination of the broad outline of the stratigraphy of these formations and the settling of their age as lower Pleistocene.

In the course of the new French-Kenyan-American-Ethiopian expedition, of which the first field season has just ended, it was possible for us to provide two new pieces of data.

A. First of all, a series of stratigraphic determinations, completing the results of 1933, have permitted the establishment, for all of the Omo beds, of a chronological succession of at least three periods:

1) a basal series with *Elephas africanavus*.

2) a middle series with *Elephas hysudricus recki*, containing the classic fauna of the Omo and constituting the most important mass of the deposits.

3) an upper series where the fauna contains certain more recent elements such as *Hippopotamus amphibius*, associated with surviving elements from the preceding level (such as *E. recki* in a more evolved, very hypsodont, form).

B. The second element is the discovery of a humanoid mandible. This piece was found in precise stratigraphic position in a sandy level of the middle stratigraphic zone, where it was associated with typical elements of that zone: *E. recki*, *Dinotherium bozasi*, *Hippopotamus protamphibius*, etc. It is an adult mandible whose

From C. Arambourg and Y. Coppens, "Découverte d'un australopithécien nouveau dans les gisements de l'Omo (Éthiopie), *South African Journal of Science*, Vol. 64, pp. 58–59 (1968). Translated by W. E. Meikle by permission of the Foundation for Research Development, Pretoria.

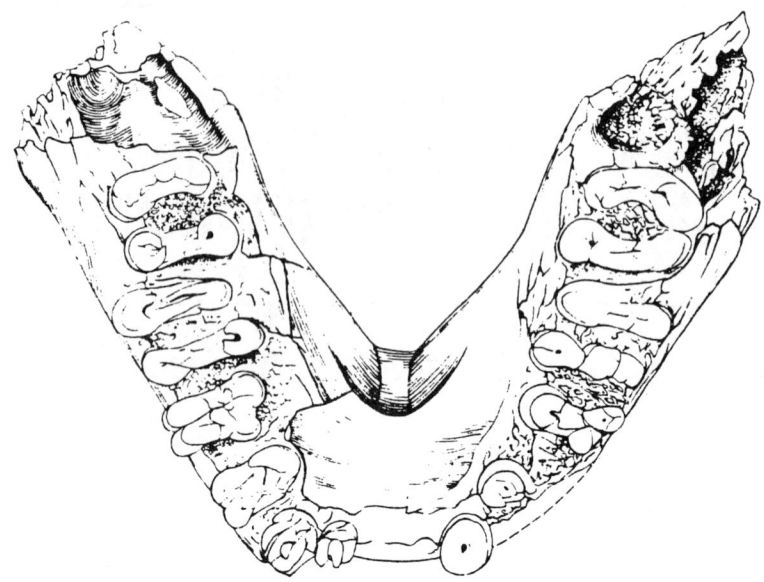

Figure 1. Mandible of *Paraustralopithecus aethiopicus*. Occlusal view. Natural size.

dental series has sadly been sheared off at the level of the cervix. However, the roots remaining in place permit us to ascertain the very weak development of the incisor series, the smallness of the canines, the general macrodonty of the cheekteeth. In other areas, the essential characteristics of this mandible are: its extreme robusticity; the thickness and low height of the horizontal rami which diverge in a "V"; the low symphysis, thick, curved and receding; the very reduced planum alveolare, slightly inclined, and the presence, immediately below this planum, of a profound genioglossal fossa provided with two foramina. Finally, the very wide digastric impressions occupy a frankly human position on the lingual face of the symphyseal region. The mental foramina occupy an equally human position approximately halfway up the corpus.

The cheektooth series includes in place only the roots of P_3-M_2 and the empty alveoli of M_3.

The principal dimensions of this specimen, in millimeters:

Total length: 75

Height of the corpus below M_2: 33

Thickness of the corpus below M_2: 26

Height of the symphysis: 35

Distance between the lingual borders of the M_2's: 42

Breadth of the incisal border, between the canines: 16.5

Length of planum alveolare: 14

Length of the C-M_2 series: 56

Mesial-distal length of M_2: 14

Buccal-lingual width of M_2: 13

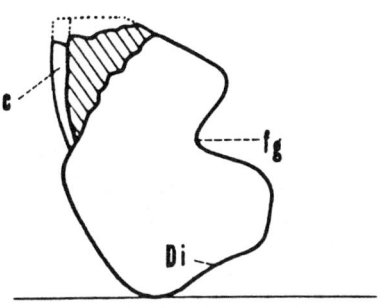

Figure 2. Section of the symphysis of *Paraustralopithecus*. C, canine; Di, impression of the digastric muscle; fg, genioglossal fossa. Natural size.

At first sight this specimen evokes, by the ensemble of its characters, the mandibles of australopithecines, particularly those of *Australopithecus africanus* from Makapansgat and of *Paranthropus*.

However, it differs nevertheless distinctly from all, and especially from *Paranthropus* (or *Zinjanthropus*), in the more stocky, low, and relatively thick form of its horizontal rami, as well as in its low, convex and very receding symphysis. Finally, it is distinguished also by the extreme reduction of its incisal region (more marked than in all other known forms), as well as of its premolars. Moreover, the reduction of the planum alveolare and its weak slope, which are the signs of a very weak freedom of the tongue, distinguish it from all the other australopithecines in evoking a structure which, in the other primates, is encountered to the same degree only in the Old World monkeys.

These diverse peculiarities suffice to show that the Omo fossil belongs to an australopithecine which is different from all other known forms and, in certain regards, is more primitive.

Its stratigraphic position, in a well defined level corresponding to the middle phase of the Villafranchian stage, seems to confirm this last observation.

We have designated this archaic form by the name *Paraustralopithecus aethiopicus*.

13

Homo ergaster Groves and Mazák, 1975
Type specimen: KNM-ER 992

In this paper Colin Groves and Vratislav Mazák propose a new species of the genus *Homo* based on a mandible and other specimens from east of Lake Turkana (referred to here as Lake Rudolf, its former name). The holotype was found in 1971 during fieldwork led by Richard Leakey and had been briefly described in the literature, as cited by Groves and Mazák. Note that the geological ages of specimens from the "lower levels" of "East Rudolf" (now known as Koobi Fora) are today known to be considerably younger than suggested in this paper: generally less than 2 million years rather than more than 2.7 million years (Feibel et al., 1989).

The name *Homo ergaster* was seldom used or cited in the years immediately following its publication. Many workers considered it a probable synonym of *Homo erectus*, to which Groves and Mazák do not compare their new species. In the 1990s, however, some paleoanthropologists have concluded that specimens formerly considered to be African *Homo erectus* represent a different, if related, species from the "classic" *Homo erectus* of Java and China. The availability of the name *Homo ergaster* has led to its adoption by some as the "new" name for what is not actually a newly-discovered species (Wood, 1991). The validity of *Homo ergaster* is related to the question of how many species of "early *Homo*" are represented in the Plio-Pleistocene fossil record of Africa and how various specimens should be allocated to those species. See Wood (1991) for a thorough review of these unresolved questions.

Colin Groves is a taxonomist, the only professional mammalian taxonomist among the authors of names represented in this book; all the others were primarily trained in other disciplines, even those like paleontologist Robert Broom who named many species during their careers. As a result, the structure of this paper reflects the formal conventions of zoological taxonomy, although its data consist largely of dental measurements taken from the published literature. Groves has continued to be interested in the hominid fossil record and has recently published a long review of hominid taxonomy (Groves, 1989).

An Approach to the Taxonomy of the Hominidae: Gracile Villafranchian Hominids of Africa

Colin P. Groves and Vratislav Mazák

1. Towards a Taxonomy of the Hominidae

Anthropologists are not, in most cases, trained in zoological modes of thought; even those with understanding of zoology tend to shy away from applying the principles involved to the *Hominidae*. This applies especially to taxonomy; most workers concerned with studies on the evolution of man have hesitated to interpret their findings using a taxonomic framework. This is in part a welcome reaction to the over-interpretation of the earlier part of this century and the latter half of the previous one: a glance at the list prepared by Campbell (1965) reveals a woeful record of misuse of nomenclature, with resultant chaos in the taxonomic field.

A consensus taxonomic scheme for the *Hominidae* would probably run somewhat as follows (most lately, for example, in Poirier, 1974):

Genus:
  Ramapithecus Lewis, 1934 species: *punjabicus* (Pilgrim, 1910)
Genus:
  Australopithecus Dart, 1925 species: *africanus* Dart, 1925
  robustus (Broom, 1938)
Genus:
  Homo Linnaeus, 1758 species: *erectus* (Dubois, 1894)
  sapiens Linnaeus, 1758.

From Colin P. Groves and Vratislav Mazák, "An Approach to the Taxonomy of the Hominidae: Gracile Villafranchian Hominids of Africa," *Časopis pro mineralogii a geologii*, Vol. 20, pp. 225–246 (1975). Reprinted by permission of *Časopis pro mineralogii a geologii*, Prague. Tables 1–7 consisting of dental measurements and data that originally accompanied this paper, along with all references to them in the text, have been omitted in this reprinted version.

This is essentially a "grade" type of classification; the "gracile" and "robust" types of Villafranchian hominids are classed together in a genus *Australopithecus* by virtue of shared primitive characters, which is no basis for a meaningful association, and the Middle Pleistocene hominids are united into a catch-all *Homo erectus*, which may be biologically correct but, as will be shown in a later paper, has caused much confusion.

Although the classification above has attained a spurious orthodoxy by reason of its repetition in numerous text-books, some reputable authorities have challenged it. Citing only text-books for the moment, we may mention Robinson (1972), who places *robustus* in a separate genus *Paranthropus* Broom, 1938 and incorporates *africanus* into *Homo*; Napier (1970), who recognizes *Paranthropus* but retains *Australopithecus*, and acknowledges a third species of *Homo*, viz. *Homo habilis* L. S. B. Leakey, Tobias & Napier, 1964; Campbell (1974), who includes *robustus* in *africanus* but recognises a second species in *Australopithecus*, known as *"Zinjanthropus" boisei* L. S. B. Leakey, 1959; and Howells (1973), who recognizes *Paranthropus* with *robustus* and *boisei*, and is cautious about *habilis*.

There has however only rarely been any attempt to quantify taxonomic differences within the *Hominidae*. Only the dispute between Robinson (1965) and Tobias (1965) over the validity of *Homo habilis* was marked by much reference to metrical data, and even here theoretical concepts such as "phyletic valence"—a concept later cogently criticised by Tobias (1967)—were permitted to get in the way of objective interpretation.

The relationship between taxonomy and phylogeny has generally remained unstated. Although for the lowest levels one looks at differences only—if two given forms differ then they have to be awarded separate taxonomic status, and if they do not then they have to be placed together—in the case of generic and familial categories it is as well to have some acknowledged theoretical standpoint: not, it is true, as to which differences should or should not be permitted to carry weight in one's conclusions, but as to what taxonomic treatment one should give to a known phylogenetic scheme. This, of course, implies that the approximate course of the group's evolution is known; as fossil remains augment, and as studies on living forms attain increasing rigour, this evolution should become better and better known so that a taxonomic framework can be constructed which will be subject to only minor adjustments with later discoveries.

It has been repeatedly shown (see, for example, Brundin 1972) that only a Hennigian phylogenetic type of classification is truly logical. Only members of a common phyletic lineage can be said to be "related" in any meaningful way; the common possession of primitive characteristics—such as defines the *Australopithecus* in the "consensus classification" quoted above—is no basis for taxonomic union unless the forms under consideration share an exclusive common ancestor. This means that, if the taxon generally known as *Australopithecus africanus* shared a common ancestor with *Homo* after the divergence of *Australopithecus robustus*, as almost all palaeoanthropologists seem to believe, then it must be placed in some taxonomic grouping with *Homo* that excludes *Australopithecus robustus*; or in other words, that the species *robustus* cannot meaningfully be placed in the genus *Australopithecus*.

The Hennigian scheme is not specially concerned with palaeontology, so says nothing about what one does with common ancestors. To the extent that such beings have not knowingly been identified in the *Hominidae*, this question is irrelevant to the problem at hand, but its existence should not be ignored. One of us has elsewhere (Groves 1974) proposed to include common ancestors in a group with the descendant lineage that has changed least, which is the basis for his adoption of the primate family *Pliopithecidae*; but other courses of action are conceivable.

Both "robust" and "gracile" australopithecines have been plausibly identified from White Sands at Omo (Howell 1969) and from below the KBS tuff at Rudolf (Brock & Isaac 1974), both these sites being above 2½ million years old; there is also (Howell l. c.) an unmistakably "gracile" tooth from Loc. 2 in the Shungura Formation at Omo (ca. 3.6 million years), implying the coexistence of the two types of australopithecines even this far back. If we therefore separate the two types into two genera, we will be consistent with other recent palaeontological works (e.g. Maglio 1973, who divides Plio-Pleistocene elephants into three genera with a 5-million-year time depth). For the robust type the name *Paranthropus* Broom, 1938, with type species *robustus* Broom, 1938, is available; for the gracile type, if with Robinson (1972) we accept that the "gracile australopithecine" (i.e. *Australopithecus s. str.*) has a close phyletic connection with the line leading to modern man, there is no other reasonable course of action but incorporation into *Homo*.

The view of Campbell (1972) that the South African "robust" types are not really distinguishable from the "gracile" types of the same region, will be discussed in a subsequent paper. For the moment it will be accepted that the two are separate, and that they occur together, under conditions of evident reproductive isolation, at Swartkrans (Clarke & Howell 1972). The legitimacy of recognising two contemporary species in East Africa is not, and never has been except on purely theoretical grounds (Wolpoff 1971) in question.

2. Method

All published dental measurements of African Plio-Pleistocene "gracile" hominids have been used for comparisons. Statistical parameters (mean, standard deviation, in some cases coefficient of variation) have been calculated, and the results compared pairwise by means of Student's *t-test*. This test indicates whether, on the given evidence, any difference between a pair of samples is likely to be real rather than just an accident of sampling, and if so, at what level of probability. Traditionally, a probability of less than $p = <0.05$ (i.e. there is a less than 5% chance that the two samples are drawn from the same population) is taken as indicating a significant difference; and this significance increases as probability decreases. Note that this procedure takes sample size into account, so that to point out that sample sizes are small and to use this as an argument that no inferences can be drawn, is not a valid criticism; in the same way, variation is taken into account so that, if a *t-test* indicates a low level of probability of the null hypothesis, "wide range of variation" cannot be used as a valid criticism.

However t indicates only whether a found difference between samples is likely to be real or not; it says nothing about the taxonomic status of the difference. Mayr, Linsley & Usinger (1953) suggest a means of deciding on whether a difference should be recognised taxonomically or not: the Coefficient of Difference. This does not take sample size into account: it assumes that a test to dispose of this source of inaccuracy has already been made. A coefficient of difference of greater than 1.27 indicates a 90% joint non-overlap, which is equivalent to the "75% rule" of subspecific differentiation.

The study presented here is concerned almost entirely with dentition. In cases where cranial or even postcranial material is available, a complete assessment has to take these into account; but the paucity of such material compared to the relative richness of dental remains gives added emphasis to the latter.

The measurements used in this study are given in the following sources: Robinson 1956; Tobias 1965; Tobias & von Koenigswald 1964; M. D. Leakey, Clarke & L. S. B. Leakey 1971; Day & R. E. F. Leakey 1973, 1974; R. E. F. Leakey & Wood 1973, 1974a, b; Howell 1969; Dart 1948; and Frayer 1973. A discussion in Tobias (1965) between Tobias and Robinson, and comparison of measurements made by Frayer (1973) with other sources, leads to a fair degree of confidence that measurements taken by different authors are quite comparable and may legitimately be compared with each other.

3. Sterkfontein and Makapansgat

There is no doubt that the hominid remains from Sterkfontein and Makapansgat, classified here in the species *Homo africanus* (Dart, 1925) (originally described from Taung as *Australopithecus africanus* Dart, 1925) are closely similar. It is a pity that the dating of the respective remains is so uncertain, but the great similarity between the two samples leads to the conclusion that either (1) both sets of remains were deposited at approximately the same geological epoch, or (2) deposition occurred, if over long spans, then over much the same time span in each case, or (3) if the two were deposited at different times or over different time spans, then no morphological differentiation had taken place in the meantime. The two samples differ significantly only in the mesiodistal diameters of their lower incisors. In that the sample size of four for Makapansgat represents, in fact, two jaws, ample scope for the differences to be caused by wear alone is left. The single measurement of a Makapansgat upper incisor is not significantly different from the mean of three from Sterkfontein, and this indicates that wear factors may be involved in the case of the lowers. Incisor dimensions, especially their mesiodistal dimensions, are the only ones along the toothrow which will change markedly with relatively little wear.

4. Taung

Partridge (1973) suggests, on geomorphological grounds, that the site of Taung whence comes the holotype of *Australopithecus africanus* Dart, 1925 may be much later in time than the Sterkfontein and Makapansgat. Tobias (1973) therefore suggests that its status needs to be reassessed; that, in fact, it is probably a "robust australopithecine": consequently he proposes to avoid nomenclatorical confusion by ousting the Taung specimen from its type status and making one of the Sterkfontein specimens a lectotype. Olson (1974) justifiably pointed out that this is simply not possible under the rules of nomenclature; with characteristic scrupulousness, Tobias (1974) retracted, and acknowledged his over-hasty proposal. He is, on his own statement, presently restudying the skull to determine its true status.

Measurements of the only permanent teeth of the Taung skull are closely comparable to those from Sterkfontein and Makapansgat. It will be recalled, too, that an exhaustive study by Robinson (1954, 1956) found only slight differences between the Taung individual and the rest; so that for the present Robinson's classification will be accepted without demur.

5. Swartkrans

In anticipation of a later paper in this series, it will be categorically stated that the small hominid named *Telanthropus capensis* Broom & Robinson, 1950 cannot rationally be accommodated in *Paranthropus*. This is true on dental grounds and on the proportions of the mandibles; the case has recently been closed by the discovery (Clarke, Howell & Brain 1970) and description (Clarke & Howell 1972) of a facial fragment showing unmistakable affinities with *Homo*. It appears that the lower dentition shows numerous striking differences from Sterkfontein, such that the two cannot possibly be accommodated in the same taxon. The name *capensis* is preoccupied in the genus *Homo* by a name given to the Boskop calvaria (Broom 1918), so that a new name will have to be found for it if it is not a member of a taxon thus far known only from East Africa; but as the remains are currently under detailed study by R. J. Clarke, of Johannesburg, his verdict must be awaited before a full taxonomic assessment is possible.

6. Olduvai

Recently, Olduvai Gorge dating has been somewhat revised (cf. Curtis & Hay 1972). The lowest tuff in Bed I, tuff I B, has been dated at approximately 1.8 million years. Bed I, though very thick, was rapidly deposited, and goes thus up only to about 1.6 million years; Bed II was also rapidly deposited at first, then more slowly above the faunal break, so that the age of the lower levels of Bed II might be some 1.5 to 1.4 million years, whilst that of the uppermost levels only 1 million or even 0.9 million years.

The Olduvai gracile hominid is the notorious *Homo habilis* L. S. B. Leakey, Tobias & Napier, 1964, subject of a needless controversy at the time of its description. It was complained that the differences from *Australopithecus africanus* had "no phyletic valence" (Robinson 1965), or that there was insufficient "morphological space"; the author of the latter criticism also pointed out (Campbell 1964) that the describers had said that *Telanthropus capensis* "may well prove, on closer comparative investigation, to belong to *Homo habilis*," and so had not shown the species effectively to have existed, as a distinct taxon—a criticism which would be perfectly justified were the name *capensis* itself valid within the genus *Homo*.

Four tolerably complete calvaria of this taxon have cranial capacities as follows (Tobias 1971; Holloway 1973):

OH 7 (type) (Bed I)   — 687 cm³ (adult value)
OH 24 (Bed I)   — 590 cm³
OH 16 (Lower Bed II) — 650 cm³
OH 13 (Upper Bed II) — 650 cm³

Mean   — 637 cm³, standard deviation 64

This compares with a mean for 6 Sterkfontein and Makapansgat individuals of 442 cm³, standard deviation 21.6. This difference alone would suffice to exclude *Homo habilis* from *Homo africanus*, but the dental evidence shows that the former differs also in its significantly narrower premolars.

It has recently been claimed (R. E. F. Leakey 1974) that one of the above quoted specimens, OH 24, is not in fact *Homo habilis* but is of the same taxon as the Sterkfontein sample, i.e. *Homo africanus*. As far as we can see, there are no facts that would justify this, except perhaps for the specimen's rather small cranial capacity. M. D. Leakey, Clarke & L. S. B. Leakey (1971) demonstrated its considerable differences from Sts 5, the most complete skull from Sterkfontein: its less marked postorbital constriction, slightly more vertical forehead, less projecting glabella, more expanded parietals, deeper mandibular fossae, backwardly extended occiput, narrow keeled nasals, non-flaring zygomata. Where there are comparable skull portions for the other three specimens, OH 24 resembles them rather than Sts 5. Indeed as the three authors show it closely resembles OH 13 in its known parts (there is probably not much time difference between them, owing to the extremely rapid deposition of the early levels at Olduvai), and differs more from OH 7 and OH 16; they favour an interpretation involving sexual dimorphism, the two former being females and the latter males.

Holloway (1973) gives as 727 cm³ the cranial capacity of OH 12, a specimen discovered some ten years ago but still undescribed. Being from Bed IV and therefore somewhat under 1 million years old, it would be expected to be big-brained and robust like contemporary hominids in Java and like the earlier OH 9 from uppermost Bed II; but it is not—it is small and gracile, closely resembling *Homo habilis* to which indeed Holloway refers it, with a query. The possibility must be borne in mind that a gracile, small-brained hominid lingered on contemporary with *Homo erectus*; the alternative view, for which there is also much to be said, is that the hyper-robust OH 9 ("*Homo leakeyi*"), figured by L. S. B. Leakey (1961), is not in fact *Homo erectus*. We thank staff of the former Centre for Prehistory and Palaeontology, Nairobi, for demonstrating both these specimens to one of us (C. P. G.)

7. East Rudolf, upper levels

Maglio (1972) has divided the East Rudolf deposits into three faunal stages and Brock & Isaac (1974) have applied tentative dates to them, allocating as firmly as possible under present circumstances—which is, in fact, surprisingly firmly—the various "faunal sets" to the three levels.

The hominid remains from these deposits include both "gracile" and "robust" forms. The gracile remains occur throughout the sequence, being most plentiful from the *Loxodonta africana* zone, dated at 1.4 to 1.6 million years ago, that is somewhat later than Bed I at Olduvai. The remains differ from Sterkfontein (*Homo africanus*) in their smaller size, especially the third molars; but the upper central incisor (in a single known upper dentition!) is much larger. A graph of dental sizes (Figure 1) shows that, in the lower dentition also, the Sterkfontein and Rudolf sizes diverge more strongly backwards along the toothrow, and Olduvai resembles Sterkfontein in this respect. The premolars lack the narrowness of those of *Homo habilis*, and all the cheekteeth except P3, in the lower

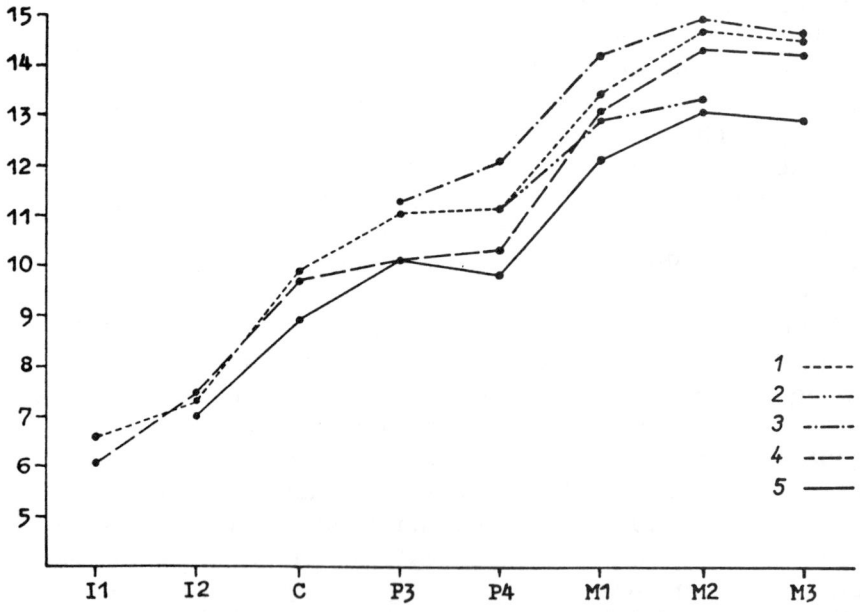

Figure 1.   Sizes of lower teeth in various South and East African Villafranchian hominids. Measurements are expressed as combined mesiodistal and buccolingual diameters (mean values), divided by two (i.e.(md + bl)/2)

1 — Sterkfontein; *2* — Omo (early); *3* — Omo (late); *4* — Olduvai; *5* — Rudolf (late)

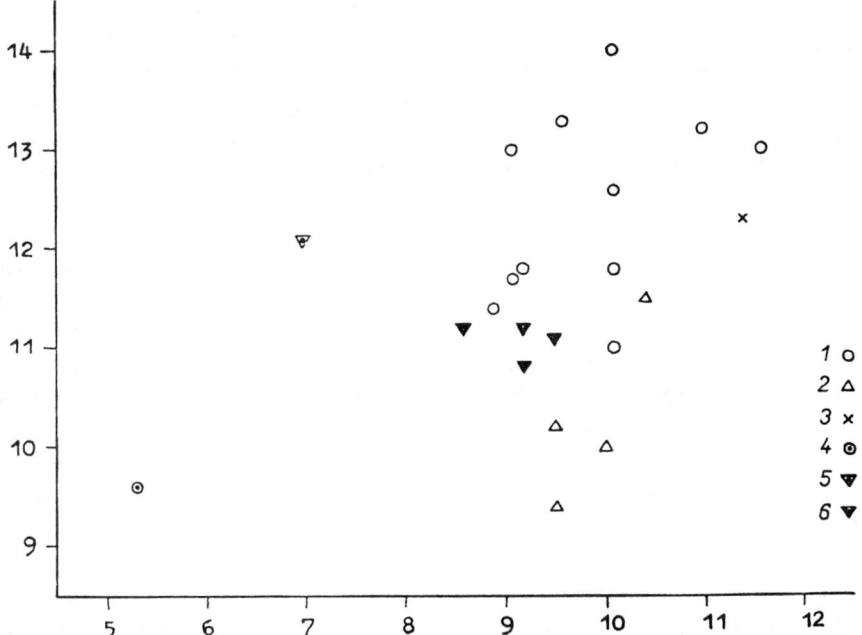

Figure 2.   Breadth-length ratio in the third lower premolars in various South and East African Villafranchian hominids. Ordinate: buccolingual diameter; abscisse: mesiodistal diameter.
1 — Sterkfontein, Makapansgat; *2* — Olduvai; *3* — Omo (late); *4* — Rudolf KNM-ER 1501; *5* — Rudolf early; *6* — Rudolf late

jaw, are smaller than the latter (Fig. 2). It thus appears that at Rudolf a different taxon again is represented.

Specimens from the earlier *Metridiochoerus andrewsi* zone (1.6–1.8 ma) at Rudolf are very similar; although contemporary with Olduvai, they differ in exactly the same way. The number of specimens in this zone is too small for adequate statistical tests. The only marked difference from the later sample is in the large size of the single known lower M3.

The breadth-length indices of the premolars, which differ so strongly between *Homo africanus* and *Homo habilis*, are much closer in the later Rudolf specimens to the former, but slightly narrower in the case of lower P3 (and that on average only).

Robinson (1956) found that the number of roots of the premolars, both upper and lower, decreases from two in *Homo africanus* to one in *Homo sapiens*; unfortunately he gave actual frequencies for the two types, and the intermediate fused condition, only in the case of the maxillary premolars. From his indications, however, it would seem that in *Homo africanus* all, or almost all, lower P4 have two separate roots, whereas some at least of P3 have two partially fused roots. It appears that the *Loxodonta africana/Metridiochoerus andrewsi* hominids commonly have a single root on both lower premolars, this representing quite an advance over *Homo africanus* from Sterkfontein. Unfortunately again, exact figures for *Homo habilis* are not available but casts of the Olduvai specimens seem to indicate that two roots and/or partially fused roots on the both lower and upper premolars are rather frequent. *Homo habilis* of Olduvai would thus emerge as a form that is, in this characteristic at least, somewhat less advanced than the East Rudolf hominid of the *Loxodonta africana* and *Metridiochoerus andrewsi* zones.

Accordingly there is no doubt but that the middle and upper levels at Rudolf contain an undescribed hominid species. R. E. F. Leakey (1974) mentions a skull, KNM-ER 1805, from these levels, with a cranial capacity between 600 and 700 cm^3 and a "small" jaw, though measurements are not given. He aligns this with other gracile hominids from the same levels, and—although in his papers with Day and Wood he had adopted a cautious attitude—assigns them to *Homo habilis*. While the cranial capacity and the general features of the calvaria would seem to be within the range of the latter, it has been shown above that the dental evidence does not admit of this interpretation.

8. East Rudolf, lower levels

The *Mesochoerus limnetes* zone at Rudolf is separated from the others by a faunal break, and is dated to 2.7 to >3.0 ma by Brock & Isaac (1974). Most of the material (R. E. F. Leakey & Wood 1974a) is referred to *Homo sp.*, but one mandible, KNM-ER 1482, from the very oldest levels, is considered to be "something completely different". The evidence for this seems fairly convincing from its description (R. E. F. Leakey & Wood 1974b), and the cranial specimens from this zone are also of two types (neither of them being *Paranthropus*).

The *Homo* specimens have rather smaller teeth than those from later levels. Among some specimens of unknown level, from areas 123 and 127 occurs a dentition, KNM-ER 1501, with exactly the same characters; as well as small size the specimens have strikingly broad premolars; the mandibular body is in KNM-ER 1501 very narrow. All in all, they are a remarkable sample—if they do indeed belong together—and quite unexpected in their morphology. In Rudolf at least, size of teeth got larger, not smaller, over time; the premolars got narrower, and in this respect *Homo habilis* may be highly evolved.

The unknown hominid, KNM-ER 1482, has been well described by R. E. F. Leakey and Wood, and it remains here to make dental comparisons. It does not have such small teeth as the unequivocal *Homo*; they are in fact rather larger (and especially, broader) than in the upper levels hominid, and could on size grounds be referred to *Homo africanus* although morphological features such as the obliquity of the premolars might forbid this. Lower P4 is not excessively broad, the mandible is however very narrow, and curiously the premolar roots are reduced in number.

R. E. F. Leakey (1973a) provisionally associates mandible KNM-ER 1482 with skull KNM-ER 1470 which he later described (1973b). He points to the broadness of the dental arch and to the evident large size and broadness of the cheekteeth, both of them features shown in KNM-ER 1470. The latter skull has been widely reported in the popular press, and shown to specialists at meetings. It has a cranial capacity of some 810 cm³, larger than any *Homo habilis*; but compared with the latter it has some undoubtedly primitive features: the face is much longer and narrower, the malar roots much lower and more buttress-like, there is no ophryonic groove, the supraorbital torus is thicker but less protruding, the face is more concave transversely, there is greater facial prognathism though less alveolar prognathism, the nasals recede, the occiput is shorter, the foramen magnum is further back. The back of the skull and the modeling of the face strongly recall Sts. 5.

R. E. F. Leakey (1974) assigns KNM-ER 1470 to *Homo habilis* and contrasts it to KNM-ER 1813, also probably from the lower member, in which the capacity is only about 500 cm³. While the previous year he had assigned KNM-ER 1482 provisionally to the same species, implying, if these two papers are taken together, that they are continuous phyletically with the remains of the upper levels, in his paper with Wood (1974b) it was KNM-ER 1482 which was excluded from the lineage, and the contemporary KNM-ER 1483 admitted. Again, there must be doubt, until the discovery of associated pieces, about which jaw goes with which skull. Jaw 1482 is less like *Homo* than jaw 1483, but skull 1813 is less like *Homo* than skull 1470 which may have features linking it to jaw 1482! We must admit that in such a situation there is no alternative but to await fresh discoveries.

9. Omo

Remains from several levels at Omo have been briefly described by Howell (1969). To maximise sample sizes, these have here been combined into three groups: Upper levels, 1.8 to 2.1 ma (Omo-75, Omo-K-7, Omo-75-S.15, Loc. 26, loc. 28–30/31), Middle levels, 2.4 to 2.6 ma (White Sands, Brown Sands, Loc. 45, Loc. 51), and Lower levels, 3.6 ma (Loc. 2).

Teeth from the Upper levels are larger than those from Sterkfontein, but narrower, so that mesiodistal diameters are greater but buccolingual are the same size (compared to Sterkfontein). Teeth from the Middle levels are similar, but smaller: this time the mesiodistal diameters are the same size as Sterkfontein, but buccolingual are less. The single tooth, lower M1, from the Lower level, is the same length as equivalent teeth from the Middle levels, but narrower still; it closely resembles KNM-ER 1502 and 1507, two of the specimens of uncertain stratigraphy from Rudolf.

The breadth-length indices of lower premolars in the Upper levels teeth are comparable to those of *Homo habilis* from Olduvai; a single lower P4 from Middle levels is even narrower.

Clearly the Omo teeth are comparable only to those from Olduvai. In the lower dentition, where both sites are tolerably well represented, the Middle levels sample is really very similar to Olduvai, which is unexpected given that the Upper levels are much closer in time (even so, slightly earlier). Figure 1 gives the impression that the Upper levels sample, though generally larger than Olduvai, has a small lower M3.

10. Taxonomic conclusions

The Olduvai sample shows several strong differences of taxonomic value from the Sterkfontein sample; the Coefficient of Difference for lower P3 is 1.40, for lower P4 is 1.35, for the cranial capacity 2.28. The difference in postcranial skeleton claimed by the describers (L. S. B. Leakey, Tobias & Napier 1964) has never actually been refuted. The skull differences described by M. D. Leakey, Clarke & L. S. B. Leakey (1971) for OH 24 seem, as far as the evidence goes, to be valid for the species as a whole. *Homo habilis* is a good species by any standards.

There are no significant differences between the Olduvai sample and remains from Lower or Middle levels at Omo. Compared to the Upper levels sample at Omo, the Coefficient of Difference for the buccolingual diameter of lower M1 is 1.14, which is near the traditional level. Some isolated teeth from this level are also unlike Olduvai: lower P3 mesiodistal is more than 3 s.d. removed from the Olduvai value, lower P4 buccolingual more than 4. These figures are suggestive, but for the moment no taxonomic differentiation can be made, and in the absence of cranial material little more can be said but that the Omo gracile hominid is probably *Homo habilis*, but may turn out to represent a different subspecies.

The most interesting series of remains, for present purposes, are those from East Rudolf. Material from the upper levels, 1.4 to 1.8 million years old, are plentiful enough to enable them to be classified with a fair degree of certainty. They are not *Homo habilis*, as the teeth are smaller (C. D. for lower M1 mesiodistal is 1.39, for lower M2 1.47 and for lower M3 even 3.16) and less narrow; they are not *Homo africanus* as the teeth are smaller (C. D. for lower M2 mesiodistal is 1.79 and for lower M3 1.97, and for lower M2 buccolingual 1.33), at least the molars, and the premolar roots are reduced; evidently the cranial capacity is larger, as in *Homo habilis* or, rather probably, even more. Creation of a new species seems perfectly justified; the discoverer, R. E. F. Leakey, has given his opinion on the taxonomic allocation of the remains (1974), and it is now open for other workers to agree or disagree, and if the latter, to erect a new taxon as required by the Code of Zoological Nomenclature.

A taxonomic scheme of African Villafranchian hominids of the genus *Homo* follows.

1. *Homo africanus* (Dart, 1925)

1925   *Australopithecus africanus* Dart, Nature, 115:195. Taung.
1936   *Australopithecus transvaalensis* Broom, Nature, 138:486. Sterkfontein.
1948   *Australopithecus prometheus* Dart, Am. J. Phys. Anthrop., 6:259. Makapansgat.

Distribution: South Africa

Horizon: unknown.

2. *Homo habilis* L. S. B. Leakey, Tobias & Napier, 1964

1964   *Homo habilis* L. S. B. Leakey, Tobias & Napier, Nature 202:8. Olduvai Gorge.

Distribution: Olduvai; Omo Valley.

Horizon: Olduvai, Bed I; Olduvai, lower and middle part of Bed II; Omo, all(?) levels. Definitely, 2.6 to 1.7 ma; perhaps (Omo) back to 3.6 ma.

3. *Homo ergaster* spec. nov.

1974   *Homo habilis*, R. E. F. Leakey, Nature, 248:656. East Rudolf.

Holotype: KNM-ER 992, two associated hemimandibles with complete dentition except for first incisors; described by R. E. F. Leakey &

Wood (1973); housed at the Kenya National Museum, Nairobi. Leakey & Wood (l.c.) and Brock & Isaac (1974) in some places give the number of the specimen as 922, but there is enough evidence that 992 is the correct registration number.

Type locality: Ileret, above the Middle Tuff, East of Lake Rudolf, Kenya.

Distribution: East Rudolf beds.

Horizon: *Metridiochoerus andrewsi* and *Loxodonta africana* zones; 1.8 to 1.4 ma.

Hypodigm: KNM-ER 730 (mandibular fragment with teeth; area 103, Koobi Fora Ridge, East Rudolf), KNM-ER 731 (mandibular fragment; Ileret, East Rudolf), KNM-ER 803, 807, 808 (maxillary dentitions and fragments; Ileret, East Rudolf), KNM-ER 806, 809 (mandibular dentitions; Ileret, East Rudolf), KNM-ER 820 (mandibular fragment with teeth; Ileret, East Rudolf), KNM-ER 992 (holotype), KNM-ER 1480 (mandibular fragment with teeth; Ileret, East Rudolf). In addition, parietal fragment KNM-ER 734 (area 103, Koobi Fora Ridge, East Rudolf) and skull KNM-ER 1805 (BBS Tuff at Karari, East Rudolf) also seem to belong here.

All the above listed specimens are housed at the Kenya National Museum, Nairobi.

Diagnosis: incisors and canines similar in size to *Homo africanus* (Dart, 1925) and *Homo habilis* L. S. B. Leakey, Tobias & Napier, 1964, but premolars and, especially, molars much smaller; premolars less narrow than in *Homo habilis* and with roots reduced in number, often only a single root is present; lower P4 seems to tend to have relatively smaller crown surface than lower P3; cranial capacity substantially larger than in *Homo africanus*, apparently as in *Homo habilis* or somewhat more; mandible rather thick and massive.

Origin of name: *Greek*, "workman." In reference to the stone tools which have now (Brock & Isaac 1974) been found in levels contemporary with the skeletal and dental remains. Under normal circumstances it would be customary to honour the discoverer (i.e. Mr. R. E. F. Leakey) by naming a new form, which he had declined to name, after him; but the name *leakeyi* is preoccupied in the genus *Homo*.

4. *Incertae sedis*

A. Swartkrans, South Africa

1950   *Telanthropus capensis* Broom & Robinson, Am. J. Phys. Anthrop., 8:155. Swartkrans. A certain member of *Homo*; dental measurements most like *Homo ergaster spec. nov.*, but too poorly known to permit accurate placement. Undated.

B. East of Lake Rudolf, Kenya

Hominids from Lower Member, below the KBS Tuff (2.6 ma and more). Two gracile hominid forms probably present.

11. Discussion

The new species *Homo ergaster*, described here for the first time, is known primarily from mandibular remains and is based mainly on the differences in mandibular dentition. All material referable to this species originates from the East Rudolf area and, as given in the previous section of this paper, was collected at various localities in the *Metridiochoerus andrewsi/Loxodonta africana* zones, dated to 1.8–1.4 million years ago. It is rather striking that, while approximately contemporary with *Homo habilis* remains from the Olduvai Gorge Bed I and Lower Bed II (aged also about 1.8–1.4 ma), the known specimens of *Homo ergaster* nonetheless differ so consistently from them. As has however been shown in sections 9 and 10 of this paper the species *Homo habilis* is also known from some localities in the Omo Valley, Ethiopia, which were definitely dated at 2.6 million years and it is not excluded that the species in question occurs even in much earlier deposits dated as far back as to 3.6 million years. It thus seems that *Homo habilis* represents a species which is of older phyletic derivation than the species *Homo ergaster*.

Our results nevertheless show that in the course of Villafranchian there were two well defined species of *Homo* in East Africa, geographically close (but evidently not sympatric), existing for perhaps several hundred thousand years alongside each other. It should be added that this conclusion is in no discrepancy with the concept of phyletic evolution as defined by Simpson in 1961 (this being not exactly identical with "anagenesis" of Rensch 1947) which is supposed to have been the basic evolutionary mode in the framework of the genus *Homo*.

Though there are important differences in morphology of teeth and structure of mandible (and perhaps in other characteristics as well) between *Homo ergaster* on one hand and *Homo africanus* and *Homo habilis* on the other, no such differences are apparent when *Homo ergaster* is compared to the so-called "*Telanthropus*" from Swartkrans. The two could therefore belong to the same taxon but more data will be however necessary to demonstrate complete identity on the species level.

As to much older material from lower levels of the East Rudolf area (dated under 2.6 ma) this is set apart from *Homo ergaster* and/or *Homo habilis* by several clearly marked features in dentition and there is a suspicion that two different forms of "gracile" hominids were present in the East Rudolf area at the latest Pliocene and the earliest Pleistocene. It is also likely that one of these two forms, especially the one represented by the famous specimen KNM-ER 1470, will be found, with its broad cheekteeth, to represent the precursor of *Homo ergaster*.

Acknowledgments: Many thanks are due to the following persons who have discussed the relevant questions with the authors, or have demonstrated specimens, but are not responsible for the views herein expressed: P. Andrews, A. Bilsborough, R. J. Clarke, G. Ll. Isaac, J. Jelinek, R. E. F. Leakey, Z. V. Špinar, A. G. Thorne, and A. Walker.

References

Brock A.-Isaac G. Ll. (1974): Paleomagnetic stratigraphy and chronology of hominid-bearing sediments East of Lake Rudolf, Kenya. — *Nature*, *247*, 344–348. London.

Broom R. (1918): The evidence afforded by the Boskop skull of a new species of primitive man (*Homo capensis*). — *Amer. Mus. N. H., Anthrop. Pap., 23*, 63. New York.

Broom R.-Robinson J. T. (1950): Man contemporaneous with the Swartkrans Ape-man. — *Amer. J. Phys. Anthrop., 8*, 151–155.

Brundin L. (1972): Evolution, Causal Biology, and Classification — *Zool. Scripta, 1*, 107–120.

Campbell B. G. (1964): Science and human evolution. — *Nature, 203*, 448–451. London.

_____ (1965): *The nomenclature of the* Hominidae. — *Occas. Pap. Roy. Anthrop. Inst. London, 22*.

_____ (1972): Conceptual progress in Physical Anthropology: Fossil man. *Ann. Rev. Anthrop., 1*, 27–54.

_____ (1974): *Human Evolution*. 2nd. ed., Aldine, Chicago.

Clarke R. J.-Howell F. C. (1972): Affinities of the Swartkrans 847 hominid cranium — *Amer. J .Phys. Anthrop., 37*, 319–335.

Clarke R. J.-Howell F. C.-Brain C. K. (1970): More evidence of an advanced hominid at Swartkrans. — *Nature, 225*, 1219–1222. London.

Curtis G. H.-Hay R. L. (1972): Further geological studies and potassium-argon dating at Olduvai Gorge and Ngorongoro Crater. — In: *Calibration of hominoid evolution*. Edited by W. W. Bishop & J. A. Miller. Scottish Academic Press, Edinburgh and Toronto.

Dart R. A. (1925): *Australopithecus africanus*: The man-ape of South Africa. — *Nature, 115*, 195–199. London.

_____ (1948): The Makapansgat proto-human *Australopithecus prometheus*. — *Amer. J. Phys. Anthrop., 6*, 259–284.

Day M. H.-Leakey R. E. F. (1973): New evidence of the genus *Homo* from East Rudolf, Kenya. I. — *Amer. J. Phys. Anthrop., 39*, 341–354.

_____ (1974): New evidence of the genus *Homo* from East Rudolf, Kenya. III. — *Amer. J. Phys. Anthrop., 41*, 367–380.

Frayer D. W. (1973): *Gigantopithecus* and its relationship to *Australopithecus*. — *Amer. J. Phys. Anthrop., 39*, 413–425.

Groves C. P. (1974): New evidence on the evolution of the apes and man. — *Věst. Ústř. Úst. geol., 49*, 1, 53–56. Praha.

Holloway R. L. (1973): New endocranial values for the East African early Hominids. — *Nature, 243*, 97–99. London.

Howell F. C. (1969): Remains of Hominidae from Pliocene/Pleistocene formations in the lower Omo basin, Ethiopia. — *Nature, 223,* 1234–1239. London.

Howells, W. W. (1973): *The Evolution of the genus* Homo. Addison-Wesley, Reading (Mass.).

Leakey L. S. B. (1959): A new fossil skull from Olduvai. — *Nature, 184,* 491–493. London.

_____ (1961): New finds at Olduvai Gorge. — *Nature, 189,* 649–650. London.

Leakey L. S. B.-Tobias P. V.-Napier J. R. (1964): A new species of the genus *Homo* from Olduvai Gorge. — *Nature, 202,* 7–9. London.

Leakey M. D.-Clarke R. J.-Leakey L. S. B. (1971): New hominid skull from Bed I, Olduvai Gorge, Tanzania. — *Nature, 226,* 223–224. London.

Leakey R. E. F. (1973a): Evidence for an advanced Plio-Pleistocene hominid from East Rudolf, Kenya. — *Nature, 242,* 447–450. London.

_____ (1973b): Further evidence of Lower Pleistocene hominids from East Rudolf, Kenya. — *Nature, 242,* 170–173. London.

_____ (1974): Further evidence of Lower Pleistocene hominids from East Rudolf, North Kenya, 1973. — *Nature, 248,* 653–656. London.

Leakey R. E. F.-Wood B. A. (1973): New evidence of the genus *Homo* from East Rudolf, Kenya. II. — *Amer. J. Phys. Anthrop., 39,* 355–368.

_____ (1974a): New evidence of the genus *Homo* from East Rudolf, Kenya. IV. — *Amer. J. Phys. Anthrop., 41,* 237–244.

_____ (1974b): A hominid mandible from East Rudolf, Kenya. — *Amer. J. Phys. Anthrop., 41,* 245–250.

Maglio V. J. (1972): Vertebrate Faunas and Chronology of Hominid-bearing sediments East of Lake Rudolf, Kenya. — *Nature, 230,* 248–249. London.

Mayr E.-Linsley E. G.-Usinger R. L. (1963): *Methods and principles of Systematic Zoology.* McGraw-Hill. New York.

Napier J. R. (1970): *Roots of Mankind.* Smithsonian Institution, Washington.

Olson T. R. (1974): Taxonomy of the Taung skull. — *Nature, 252,* 85. London.

Partridge T. C. (1973): Geomorphological dating of cave openings at Makapansgat, Sterkfontein, Swartkrans and Taung. — *Nature, 246,* 75–79. London.

Poirier F. E. (1974): *In search of ourselves.* Burgess, Minneapolis, Minn.

Rensch B. (1947): *Neuere Probleme der Abstammungslehre, die transspezifische Evolution.* Enke. Stuttgart.

Robinson J. T. (1954): The genera and species of the Australopithecines. — *Amer. J. Phys. Anthrop., 12,* 181–200.

_____ (1956): The dentition of the Australopithecinae. *Transvaal Mus. Mem., 9.*

_____ (1965): Comment on Tobias. — *Curr. Anthrop., 6,* 403–406.

_____ (1972): *Early hominid posture and locomotion.* Chicago Univ. Press.

Simpson G. G. (1961): *The Major Features of Evolution,* 3rd ed. Columbia Univ. Press, New York.

Tobias P. V. (1965): New discoveries in Tanganyika: their bearing on hominid evolution. — *Curr. Anthrop., 6,* 391–400.

Tobias P. V. (1971): *The brain in hominid evolution.* Columbia Univ. Press. New York.

———— (1973). Implications of the new age estimates of the early South African Hominids. — *Nature, 246,* 79–83. London.

———— (1974): Reply to Olson. — *Nature, 252,* 85–86. London.

Tobias P. V.-von Koenigswald G. H. R. (1964): A comparison between the Olduvai hominines and those of Java. — *Nature, 204,* 515–518. London.

Wolpoff M. H. (1971): Competitive exclusion among Lower Pleistocene hominids: the single species hypothesis. — *Man* N. S., *6,* 601–614.

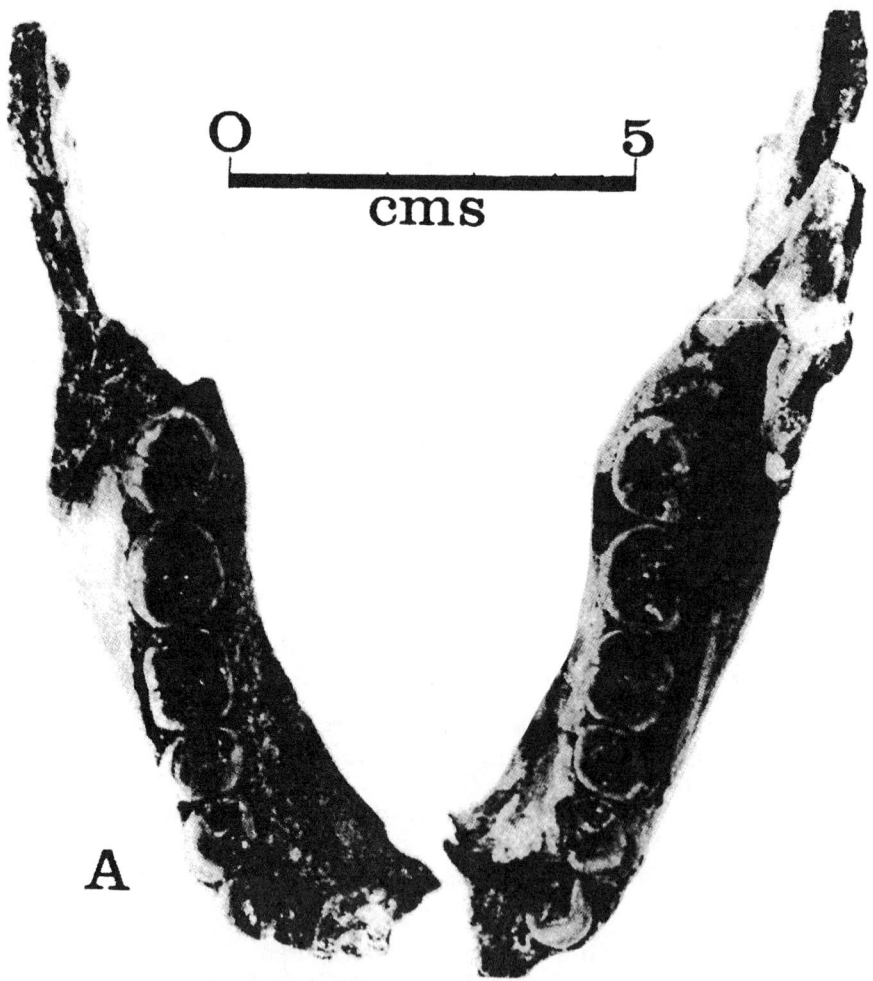

Plates I and II: Occlusal (*A*), left lateral (*B*) and left medical (*C*) views of mandible KNM-ER 992,, the holotype of *Homo ergaster* spec. nov. (After R. E. F. Leakey & B. A. Wood 1973)

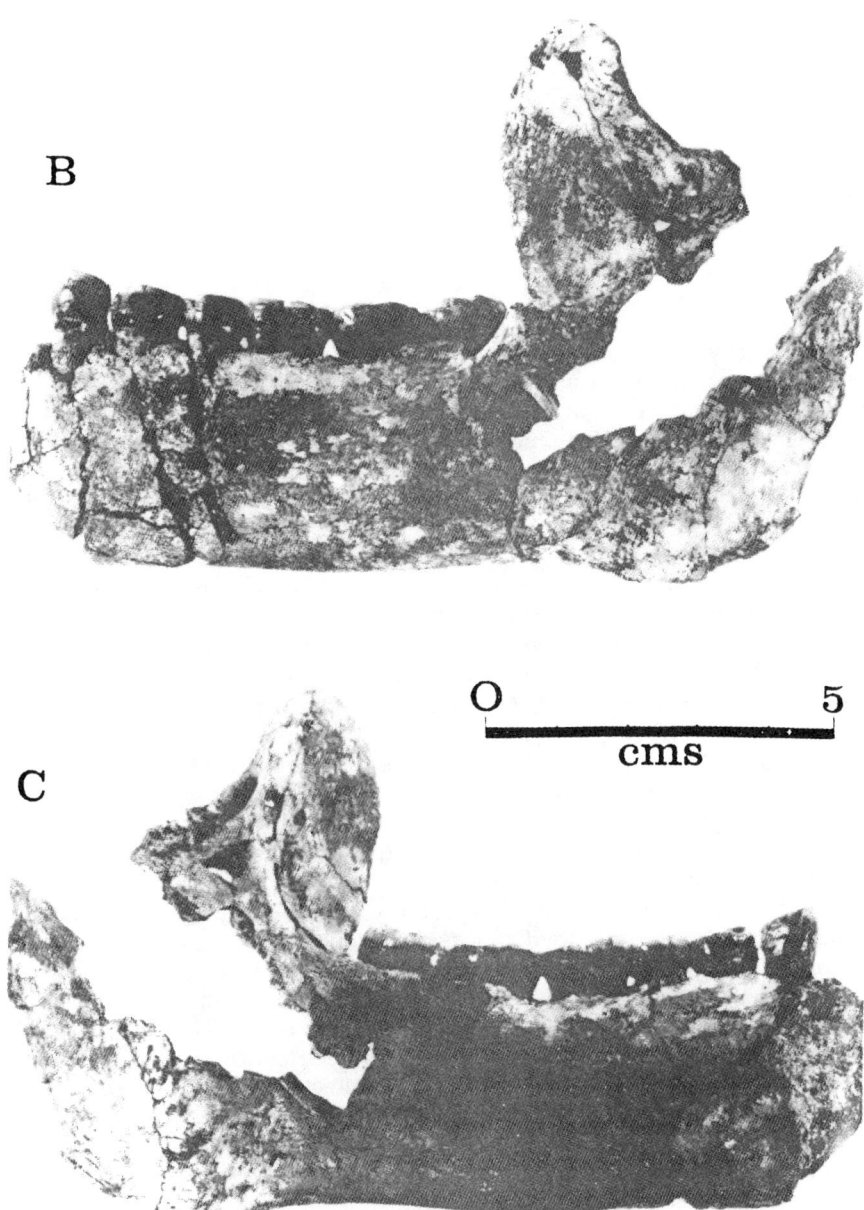

14

Australopithecus afarensis
Type specimen: LH 4

Controversy about the Pliocene hominids from Laetoli and Hadar (how many taxa are present in the sample? what is the functional significance of their post-cranial anatomy?) has continued through the two decades since large numbers of fossils were first recovered at these sites. As mentioned above, the first hominid fossils from Laetoli were collected in the 1930s (see selection 9), but only in the 1970s were the geological age and great significance of this site established (Leakey and Harris, 1987). For detailed descriptions of the Hadar fossils see the April, 1982 issue of the *American Journal of Physical Anthropology.*

Controversy over *Australopithecus afarensis* extends into nomenclature as well, as indicated by our omission of authorship and publication date above. The paper reprinted here from *Kirtlandia* was intended to formally create the new species name "*afarensis.*" Unfortunately, while it was in press a brief account of a scientific conference in Sweden appeared in *New Scientist* (Hinrichsen, 1978), in which the name *Australopithecus afarensis* was printed. This account reported in just a few sentences Donald Johanson's suggestion at the meeting of a new species. Because the *New Scientist* article appeared before the *Kirtlandia* paper, several taxonomists have concluded that the nomenclaturally-correct citation of the name of this species should be "*Australopithecus afarensis* Johanson (in Hinrichsen), 1978," rather than "*Australopithecus afarensis* Johanson, White and Coppens, 1978" (Day et al., 1980; Groves, 1989). This has been contested by Johanson and White (1980) who do not believe that Hinrichsen's article is sufficient to make this species name available. We have chosen to reprint here only the *Kirtlandia* paper because all the other papers in this book were written and intended as scientific communications among specialists. Hinrichsen's article is not; it is simply a news report of a current event. Only the fact that Hinrichsen printed the new name before the *Kirtlandia* paper was published created this situation; if *New Scientist* had omitted the name or published it later there would have been no dispute about the citation. The Code suggests (Recommendation 50A) that reporters not publish new scientific names in reports of meetings, but has no power of regulation in this matter. We reprint the *Kirtlandia* paper because it is a formal communication of taxonomic conclusions which was completed, and in press, before Hinrichsen's account appeared.

It was pointed out above (see selection 9) that the species name *Meganthropus africanus* Weinert, 1950 is based on the Garusi maxilla, which was found in the Laetoli deposits in 1939. Note that Johanson, White, and Coppens include this specimen in the hypodigm of *Australopithecus afarensis*, accepting that it is part of the same species as other specimens from Laetoli. As mentioned above, Johanson, White, and Coppens did not technically suggest a totally new species; they did vastly expand knowledge about a previously very poorly known and widely ignored taxon, while providing it with a new species name for use in the genus *Australopithecus*.

Note to the reader: A large part of this article consists of a catalogue of specimens which contains a large number of anatomical terms and abbreviations. The abbreviations fall into three categories:

1) museum catalogue designations: for example, A.L. 128-1 = Afar Locality number 128, specimen 1; L.H. -1 = Laetolil Hominid specimen 1

2) dental designations: for example, $I^2 — M^1$ = 2nd upper incisor to first upper molar (I = incisor; C = canine; P = premolar; M = molar; dm = deciduous molar; superscripts = upper jaw dentition; subscripts = lower jaw dentition; L = left side of jaw; R = right side of jaw) (Also note that P3 and P4 designate the first and second premolars; the primitive P1 and P2 were lost in primate evolution.)

3) anatomical designations: for example, prox. = proximal (or the end of a bone nearest the trunk); med. = medial (or toward the midline)

The anatomical terms fall into two categories: names of bones (calcaneus or heel bone, for example) and orientation terms (such as proximal vs. distal, or medial vs. lateral). These can be found in any anatomy book.

Also note a few taxonomic abbreviations used by the authors of this paper. The first proposal of the new species name is marked as "sp. nov." meaning new species. The designation "sp. indet." stands for species indeterminate. The term "aff." (with affinity to) means that a species is considered related to, but not necessarily identical with, another.

A New Species of the Genus *Australopithecus* (Primates: Hominidae) from the Pliocene of Eastern Africa

Donald C. Johanson, Tim D. White and Yves Coppens

A substantial collection of hominid fossils has recently been recovered from two Pliocene sites in eastern Africa. Hominid specimens from Hadar in Ethiopia (11°N, 40°30'E) and Laetolil in Tanzania (3°12'S, 35°11'E) have been dated to between ca. 2.9 and ca. 3.7 million years before present (Aronson et al., 1977; Leakey et al., 1977). The strong morphological continuity between these two samples suggests that they are best considered as representing a single taxon; hence, the Hadar and Laetolil fossils currently constitute the oldest indisputable evidence of the family Hominidae.

Some of these specimens have been provisionally allocated to *Homo* sp. indet. (Johanson and Taieb, 1976; Leakey et al., 1977) while others have been referred to *Australopithecus* aff. *africanus* (Johanson and Taieb, 1976). Subsequent to this preliminary assessment, more detailed study of the entire hominid sample from Laetolil and Hadar has provided us with new information indicating that 1) the specimens belong to only a single taxon, and 2) they differ significantly from previously recognized species of Plio/Pleistocene Hominidae. The Hadar and Laetolil hominids exhibit many morphological features found in specimens attributed to the genus *Australopithecus (sensu lato)* (as defined by Le Gros Clark, 1955) and they are therefore assigned to this taxon. Careful evaluation of the material had led to the recognition of a distinctive suite of morphological traits distinguishing the Laetolil and Hadar remains from other hominid taxa. Such study indicates the necessity of assigning these fossils to a new and more primitive species of *Australopithecus*.

From Donald C. Johanson, Tim D. White and Yves Coppens. "A New Species of the Genus *Australopithecus* (Primates: Hominidae) from the Pliocene of Eastern Africa." *Kirtlandia*, No., 28, pp. 1–14 (1978). Reprinted by permission of The Cleveland Museum of Natural History.

Order PRIMATES Linnaeus 1758
  Superfamily HOMINOIDEA Simpson 1931
    Family HOMINIDAE (Le Gros Clark 1955)
      Genus *Australopithecus* Dart 1925

Australopithecus afarensis sp. nov.
Synonomy:

1950 *Meganthropus africanus* Weinert, H.: 139
1955 *Praeanthropus africanus* Senyürek, M.: 33

Holotype: Laetolil Hominid (L.H.-)4, mandibular corpus with broken RC, M_1, M_2; intact R and LP_4; RP_3, M_3; LM_1, M_2.

Locality: Locality 7 of the Laetolil Site, Tanzania, collected in 1974 by M. Muluila.

Horizon: Laetolil Beds between Aeolian Tuffs b and c, Pliocene age (3.6–3.7 m.y. b.p.)

Paratypes:

Laetolil Beds, Tanzania:

L.H.-1, RP^4; L.H.-2, immature mandibular corpus with permanent and deciduous teeth; L.H.-3(a-t), isolated upper and lower deciduous and permanent teeth; L.H.-3/6 a, b, Rdc-, Ldm^1; L.H.-5, R. maxillary row, I^2-M^1; L.H.-6(a-e), isolated permanent and deciduous upper teeth; L.H.-7, RM^- frag.; L.H.-8, RM^2, RM^3, L.H.-10, L. edentulous mandibular frag.; L.H.-11, $LM^{1/2}$; L.H.-12, $LM^{2/3}$ frag.; L.H.-13, R. edentulous mandibular corpus frag.; L.H.-14(a-h), isolated lower teeth; Garusi maxilla, $RP^3 - P^4$.

Hadar Formation, Ethiopia:
Sidi Hakoma Member:

A.L. 128–1, L. prox. femur frag.; A.L. 128–23, R. mandibular corpus, C-M_2; A.L. 129–1a, b, c, femur and tibia frags.; A.L. 129–52, L. ischium; A.L. 137–48a, R. distal humerus; A.L. 137–48b, R. distal ulna; A.L. 145–35, L. mandibular corpus, P_3-M_2; A.L. 166–9, L. temporal frag.; A.L. 198–1, L. mandibular corpus, C-M_3; A.L. 198–17a, b, LI^1, LI^2; A.L. 198–18, RI_2; A.L. 199–1, R. maxilla, C-M^3; A.L. 200–1a, maxilla, complete dentition; A.L. 200–1b, RM_1; A.L. 211–1, R. prox. femur frag.; A.L. 228–1, R. diaphysis femur; A.L. 266–1, mandibular corpus, LP_3-M_1, RP_3-M_3; A.L. 277–1, L. mandibular corpus, C-M_2; A.L. 311–1, L.

mandibular corpus, P_3; A.L. 322–1, L. distal humerus; A.L. 400–1a, mandibular corpus, LI_1-M_3, RI_2-M_3; A.L. 400–1b, RC $^-$; A.L. 411–1, R. mandibular corpus, M_1-M_3.

Denan Dora Member:

A.L. 161–40, LM³; A.L. 188–1, R. mandibular corpus, M_2-M_3; A.L. 207–13, L. mandibular corpus; P_3-M_2; A.L. 241–14, LM _ ; A.L. 366–1, LM_3; A.L. 388–1, LM³.

A.L. 333–1, facial frag. and maxilla, RP³-P⁴, LC-P³; -2, maxilla, RC-M¹, LI²-P⁴; -3, R. prox. femur; -4, R. distal femur; -5, L. prox. tibia frag.; -6, L. distal tibia; -7, L. distal tibia; -8, R. calcaneum frag.; -9a, -9b, R. and L. distal fibulae; -10, L. mandibular corpus frag., P3; -11, R. prox. ulna frag.; -12, R. distal ulna; -13, L. prox. V metatarsal (MT); -14, R. V metacarpal (MC); -15, L. prox. II MC; -16, L. III MC; -17, R. distal V MC; -18, R. distal IV MC; -19, prox. hand phalanx; -20, prox. hand phalanx frag.; -21, distal MT; -22, prox. hand phalanx frag.; -23, cranial frag.; -25, intermed. hand phalanx; -26, prox. phalanx; -27, R. distal II MC; -28, R. medial cuneiform frag.; -29, L. distal humerus; -30, Rdm_1; -31, prox. hand phalanx frag.; -32, intermed. hand phalanx; -33, prox. hand phalanx; -34, immature metapodial; -35, Rdc _ ; -36, R. foot navicular; -37, R. calcaneum frag.; -38, L. immature distal ulna; -39, L. immature prox. tibia; -40, R. capitate; -41, R. med. femoral condyle frag.; -42, L. prox. tibia; -43a, b, L. and R. mandibular corpi, R. and L. dm_1-dm_2; -44, LM $^-$; -45, partial cranium; -46, intermed. hand phalanx; -47, R. foot navicular; -48, L. II MC; -49, prox. hand phalanx frag.; -50, R. hamate; -51, body thoracic vertebra; -52, frag. M _ ; -53, thoracic vertebra frag.; -54, L. prox. I MT; -55, L. calcaneum frag.; -56, L. IV MC; -57, prox. hand phalanx; -58, R. prox I MC frag.; -59, R. mandibular corpus frag., M_2-M_3; -60, prox. phalanx; -61, L. distal femur frag.; -62, prox. hand phalanx; -63, prox. hand phalanx; -64, intermed. hand phalanx; -65, R. prox. III MC; -66, Ldc $^-$; -67, Rdi^2; -68, Ldi_2; -69, L. prox. hand phalanx; -70, immature metapodial; -71, prox. foot phalanx; -72,

MT frag.; -73, body lumbar vertebra; -74, L. mandibular corpus frag., M_1-M_3; -75, head R. talus; -76, Ldi_2; -77, Ldc _ ; -78, L. prox V MT; -79, L. lateral cuneiform; -80, R. trapezium; -81, body immature thoracic vertebra; -82 LI^1; -83 atlas vertebra frag.; -84, R. temporal frag; -85, L. distal fibula; -86, maxilla, L. and R. dm^1-dm^2, M^1; -87, L. prox. humerus frag.; -88, intermed. hand phalanx; -89, L. V MC; -90, LC _ ; -92; immature long bone frag.; -93, prox. hand phalanx; -94, L. clavicle frag.; -95, R. prox. femur; -96, L. distal tibia; -97, L. mandibular corpus frag.; -98, R. prox. radius; -99, Ldc $^-$; -100, L. coronoid process mandible; -101, axis vertebra; -102, prox. hand phalanx; -103, RC _ ; -104, Rdc $^-$; -105, partial immature cranium, Rdm^1-dm^2; -106, cervical vertebra; -107, R. prox. humerus; -108, L. ascending ramus; -109, humerus shaft frag.; -110, L. immature distal femur frag.; -111, R. immature distal femur frag.; -113, immature long bone frag.; -115, associated foot bones.

A.L. 333w-la-e, R. and L. mandibular corpi, $LP_3 - M_2$, $RP_3 - M_2$, RM_3 frag., R. condyle; -2, LC $^-$; -3, LI_2; -4, prox. hand phalanx; -5, R. distal II MC; -6, R. prox. III MC; -7, immature prox. hand phalanx; -8, vertebra frag.; -9a, b, LI_1, LI_2; -10, RC _ ; -11, distal hand phalanx; -12, R. mandibular corpus frag., M_1; -13, prox. fibula; -15, R. coronoid process mandible; -16, L. mandibular condyle; -17, -18, -19, rib frags.; -20, immature prox. hand phalanx frag.; -21, immature phalanx; -22, R. distal humerus frag.; -23, R. immature II MC; -25, prox. hand phalanx frag.; -26, L. prox. V MC; -27, L. mandibular corpus, M_2; -28, RI^2; -29, immature prox. phalanx frag.; -30, rib; -31, L. distal humerus frag.; -32, R. mandibular corpus, M_3; -33, R. prox. radius; -34, intermed. hand phalanx frag.; -35, L. prox. V MC; -36, L. prox. ulna; -37, L. distal fibula; -38, intermed. hand phalanx; -39, R. I MC; -40, R. prox. femur frag.; -41, rib frag.; -42, RP^4; -43, immature prox. I MT frag.; -45, rib frag.; -46, R. mandibular corpus, P_3; -47, rib frag.; -48, RM_2; -50, distal hand phalanx; -51, prox. hand phalanx frag.; -52,

L. mandibular condyle; -53, intermed. hand phalanx frag.; -54, prox. hand phalanx frag.; -55, MT frag.; -56, R. distal femur; -57, L. mandibular corpus, M_2-M_3; -58, mandibular corpus frag., LI_1-P_4, RI_1-C _ ; −59, L. mandibular corpus, M_2-M_3; -60, mandibular corpus, LP_3-M_3, RI_1-C _ .

A.L. 333x-1, RM^3; -2, LI^2; -3, LC −; -4, RI^1; -5, R. prox. ulna; -6, -9, R. clavicle frags.; -12, thoracic vertebra; -13a, prox. hand phalanx; -13b, intermed. hand phalanx; -14, -15, prox. radial epiphyses; -17, RI^2; -18, intermed. hand phalanx; -20, RI^1; -21a, b, intermed. hand phalanges; -25 $di_{1/2}$; -26, R. prox. tibia.

Kada Hadar Member:
A.L. 288-1, partial skeleton.

Horizon: Laetolil Beds, Tanzania. Known hominid sample from between strata dated to 3.59 and 3.77 m.y.

Hadar Formation, Ethiopia. Sidi Hakoma Member dated to older than ca. 3.0 m.y., but less than ca. 3.3 m.y. Denan Dora and Kada Hadar Members dated to younger than ca. 3.0 and older than ca. 2.6 m.y. with the latter member stratigraphically above the former.

Diagnosis:

A species of *Australopithecus* distinguished by the following characters:

Dentition

Upper central incisors relatively and absolutely large; upper central and diminutive lateral incisors with strong lingual basal tubercles, upper incisors with flexed roots; strong variation in canine size, canines asymmetric, lowers with strong lingual ridge, uppers usually with exposed dentine strip along distal edge when worn; P_3 occlusal outline elongate oval in shape with main axis mesiobuccal to distolingual at 45°–60° to tooth row, dominant mesiodistally-elongate buccal cusp, small lingual cusp often expressed only as inflated lingual ridge; disatemata often present between I^2/C − and C _ /P_3; C −/P_3 complex not functionally analogous to pongid condition.

Mandible

Ascending ramus broad, not high, corpus of larger specimens relatively deep anteriorly and hollowed in region of low mental foramen which usually opens anterosuperiorly; moderate superior transverse torus; low rounded inferior transverse torus; anterior corpus rounded and bulbous; strong posterior angulation of symphyseal axis; postcanine teeth aligned in straight rows; arcade tends to be sub-rectangular, smaller mandibles with relatively narrow incisor region.

Cranium

Strong alveolar prognathism with convex clivus; palate shallow, especially anteriorly; dental arcade long, narrow, straight sided; facial skeleton exhibiting large, pillar-like canine juga separated from zygomatic processes by deep hollows, large zygomatic processes located above P^4/M^1 and oriented at right angles to tooth row with inferior margins flared anteriorly and laterally; occipital region characterized by compound temporal/ nuchal crest (in larger specimens), concave nuchal plane short anteroposteriorly; large, flattened mastoids; shallow mandibular fossae with weak articular eminences placed only partly under braincase; occipital condyles with strong ventral angulation.

Postcranium

See remarks.

Description:

Dentition

Large canines project beyond tooth rows and possess massive, long roots; buccal face of P_3's often with vertical wear striae caused by occlusion with upper canines; P_3's often with two distinct roots, the mesial one round and angulated mesiobucally, the distal one plate-like and oriented transverse to the tooth row; P^3's sometimes three rooted, with pointed buccal cusp, extensive and asymmetric buccal face, buccal cerviocoenamel line projecting towards mesio-buccal root, and the lingual cusp situated mesial to buccal cusp, P^3's tend to be larger than P^4's and the latter do not show mesiodistal elongation of the buccal crown half; lower molars, especially M_1 and M_2 tend to be square with cusps arranged in Y-5 pattern; wide occlusal foveae on all molars; strong molar size gradient of M3 > M2 > M1; hypocones and hypoconulids large; deciduous canines similar to the permanent ones in form and occlusal projection; dm_1's molarized, with lingually facing anterior foveae and deep buccal grooves; substantial variation in tooth size.

Mandible

Ascending ramus slopes posteriorly and joins corpus at high position defining narrow extramolar sulcus; broad condyles; mandibular canal immediately below distal M_3 root; base of corpus everted.

Cranium

Incisors procumbent; lower margin of pyriform aperture marked laterally by raised borders; tooth rows tend to converge posteriorly; strong muscle markings on vault and cranial base, temporal lines converge anteriorly, but presence of sagittal cresting unknown; lateral portion of cranial base highly pneumastised; occipital condyles placed below external auditory meatus in lateral view; estimated cranial capacity small relative to *Homo* sp.; broad mandibular fossae, laterally projecting postglenoid proces; pyramid process angles anteriorly relative to more transverse tympanic plate.

Postcranium

Strong dimorphism in body size; all skeletal elements with high level of robusticity in muscle and tendon insertions; pelvic region and lower limbs indicate adaptation to bipedal locomotion: "waisted" appearance of capitate; third metacarpal lacking styloid process; phalanges strongly longitudinally curved; foot navicular with cuboideonavicular facet; deep peroneal grooves on distal fibulae; anterior margin of ilium between anterior superior and inferior spines relatively straight; cervical vertebrae with long spinous process; relatively high humerofemoral index compared to modern humans.

Etymology:

> The species name *afarensis* derives from the Afar depression of northeastern Ethiopia, where the largest portion of the paratype series was recovered.

Remarks

Laetolil Hominid-4 was selected as the holotype both because of its distinctive, diagnostic morphology and because it has previously been fully described and illustrated (White, 1977). The generic name *Praeanthropus* originally proposed by Hennig (1948) is invalid because no species designation was given. Senyürek (1955) used the generic nomen *Praeanthropus* and utilized Weinert's (1950) specific name *africanus*, designating the original Garusi maxillary fragment as *Praeanthropus africanus*. The present authors do not consider the original Garusi maxillary fragment or the new Laetolil and Hadar material to represent a hominid genus distinct from *Australopithecus*.

The authors recognize that individual traits and even single specimens in the new collections can be matched in other samples representing different taxa (e.g., *Australopithecus africanus* Dart 1925, *Homo habilis* Leakey, Tobias and Napier 1964). However, the overall character complex seen in the Hadar and Laetolil fossils is distinct from other previously found and described species. Care has been taken in the diagnosis to follow Mayr's suggestion to "list the most important characters or character combinations that are peculiar to the given taxon and by which it can be differentiated from other similar or closely related ones" (1969: p. 266). In the description of *Australopithecus afarensis* we have chosen to present a characterization of the entire hypodigm. This should insure that the presentation not be viewed as typological and should also give some indication of the variation recognized in this new taxon.

It is important to recognize that certain traits or complexes were not considered in the diagnosis but placed in the description due to the lack of comparable anatomical specimens from other species of *Australopithecus*. Some of the traits, such as the morphology of the hand and foot bones, may be diagnostic of the new species, but this cannot be ascertained until pertinent new material is recovered from other sites.

The Hadar and Laetolil fossils appear to represent a distinctive early hominid form characterized by substantial size variation which is interpreted as reflecting sexual dimorphism. Members of this new taxon display a complex of primitive dental, cranial, and possibly postcranial characteristics. Recognition of the new species *Australopithecus afarensis* has important implications for interpretations of early hominid phylogeny. These implications will be considered in forthcoming publications.

Acknowledgments

Fieldwork at Hadar and Laetolil was undertaken with the kind permission and cooperation of the Provisional Military Government of Socialist Ethiopia and the United Republic of Tanzania respectively. We thank the following institutions for financial support: the National Geographic Society, the National Science Foundation, the L. S. B. Leakey Foundation, the Wenner-Gren Foundation, the Cleveland Museum of Natural History, the Centre National de la Recherche Scientifique, and the Singer-Polignac Foundation.

We thank Dr. Owen Lovejoy for helpful comments and Mr. Anson Laufer and Mr. Bruce Frumker for assistance with the photographs. Thanks are also due Mr. William H. Kimbel and Mr. B. Thomas Gray for providing invaluable comments, suggestions, as well as editorial and photographic assistance.

Appreciation is expressed to Dr. Maurice Taieb for his role in the discovery of Hadar and for his initiation and successful development of the International Afar Research Expedition. Special thanks are extended to Professor Ernst Mayr for critically reviewing the manuscript.

References

Aronson, J. A., T. J. Schmitt, R. C. Walter, M. Taieb, J. J. Tiercelin, D. C. Johanson, C. W. Naeser and A. E. M. Nairn, 1977, New geochronologic and palaeomagnetic data for the hominid-bearing Hadar Formation of Ethiopia, *Nature, 267*: 23–327.

Hennig, E., 1948, Quartärfaunen und Urgeschichte Ostafrikas, *Naturwiss. Rdsch. Jahrg, 1,* 5: 212–217.

Johanson, D. C. and M. Taieb, 1976, Plio-Pleistocene hominid discoveries in Hadar, Ethiopia, *Nature, 260*: 293–297.

Leakey, M. D., R. L. Hay, G. H. Curtis, R. E. Drake, M. K. Jackes and T. D. White, 1977, Fossil hominids from the Laetolil Beds, *Nature, 262*: 460–466.

Le Gros Clark, W. E., 1955, *The fossil evidence for human evolution*, Chicago: The University of Chicago Press.

Mayr, E., 1969, *Principles of systematic zoology*, New York: McGraw-Hill Book Company.

Senyürek, M., 1955, A note on the teeth of *Meganthropus africanus* Weinert from Tanganyika Territory, *Belleten* (Ankara), *19*: 1–54.

Weinert, H., 1950, Über die Neuen Vor-und Frühmenschenfunde aus Afrika, Java, China und Frankreich, *Zeit. Morph. Anthrop., 42,* 113–148.

White, T. D., 1977, New fossil hominids from Laetolil, Tanzania. *Am. J. Phys. Anthrop., 46*: 197–230.

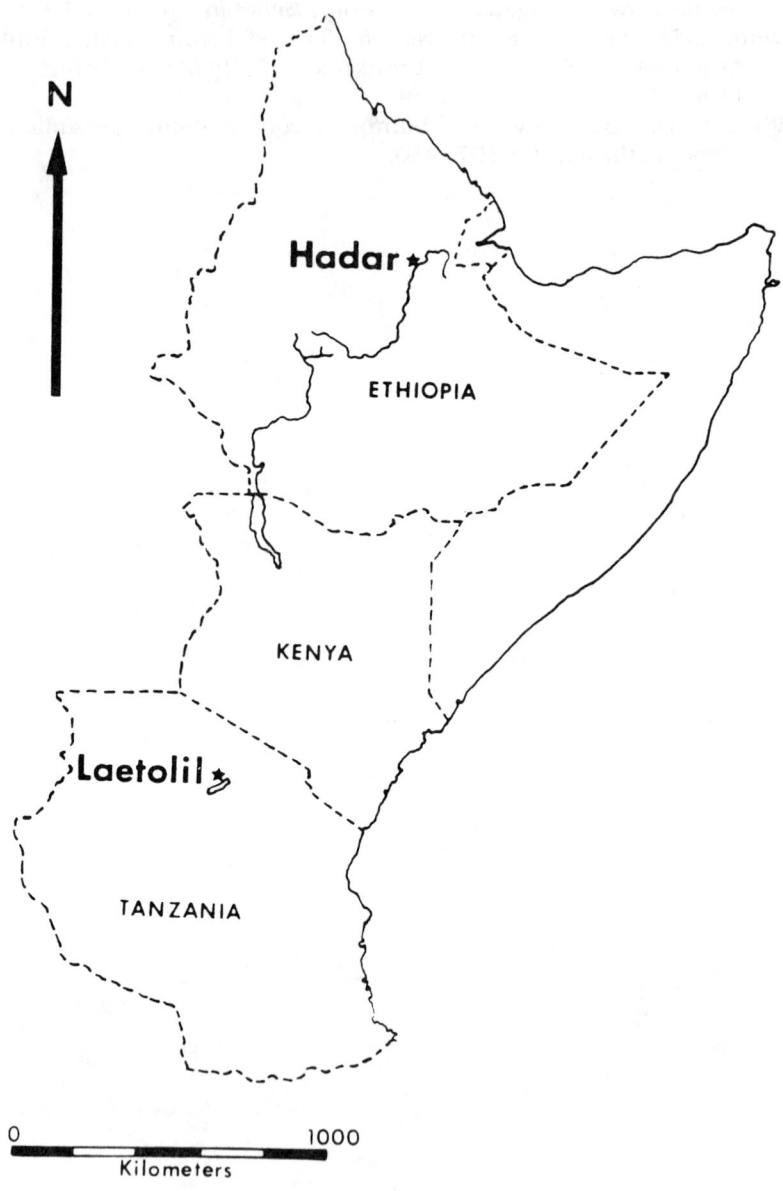

Figure 1.   Map of eastern Africa showing the locations of Hadar and Laetolil.

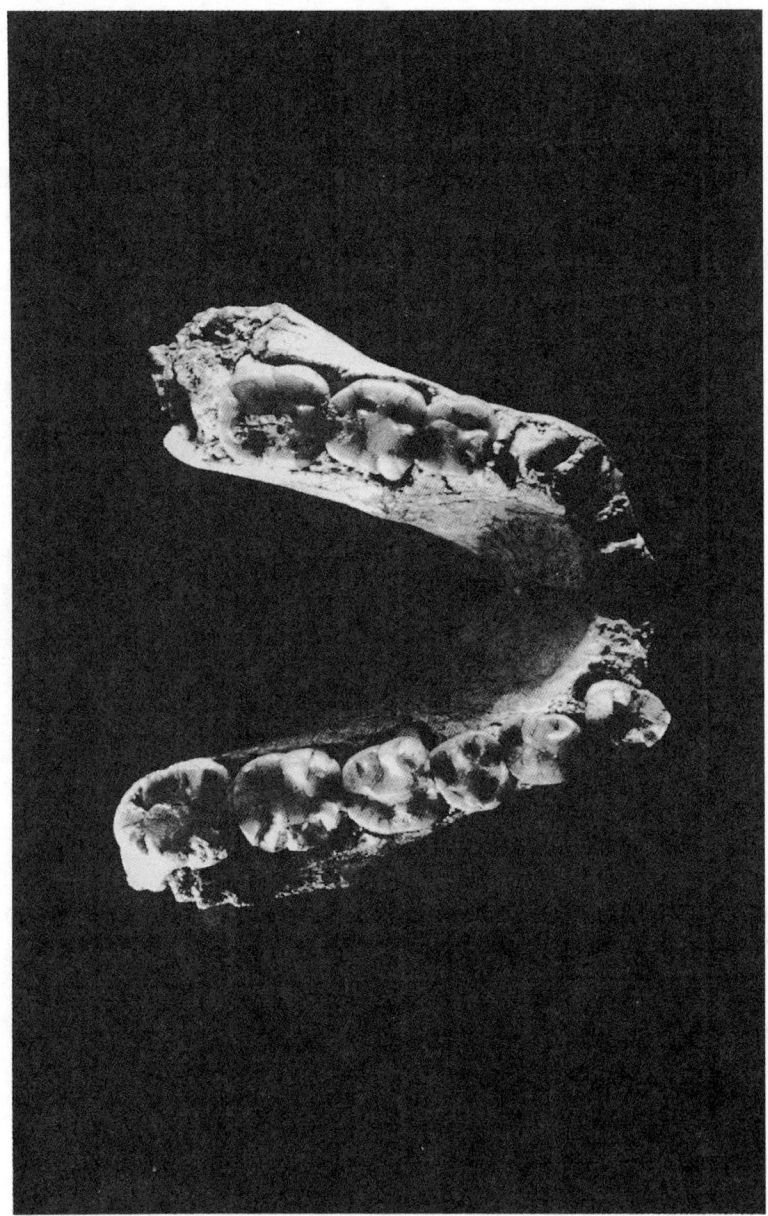

Figure 2.   Type specimen of the new species *Australopithecus afarensis*, the mandible L.H. – 4 from Laetolil, Tanzania. Occlusal view. Natural size.

Figure 3.   Two distal femora from Hadar, Ethiopia (A.L. 333—4, left; A.L. 129-Ia, right) indicating the size variation within the new species. Anterior view.

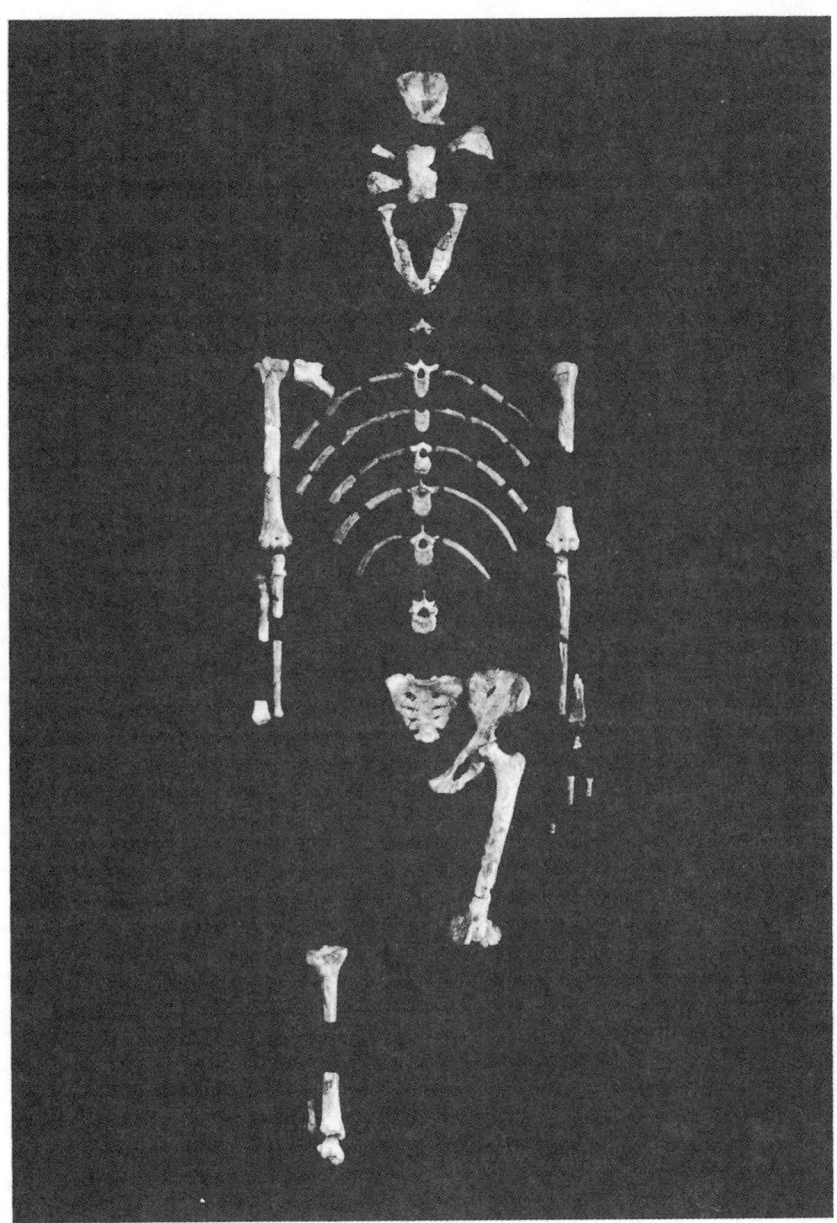

Figure 4.   The partial skeleton from Hadar, Ethiopia A.L. 288-1. The total length of the left femur is approximately 280 mm.

15

Pithecanthropus rudolfensis Alexeev, 1986
Type specimen: KNM-ER 1470

This short excerpt from V. P. Alexeev's book consists of the section where he names a new species, *Pithecanthropus rudolfensis*, with the KNM-ER 1470 cranium as its type specimen. In recent decades almost all paleoanthropologists have agreed that *Pithecanthropus* is a synonym of *Homo*, but this has no effect on the nomenclatural status of Alexeev's species name, *rudolfensis*, which depends only on its type specimen, ER 1470, and on which other fossils are grouped into a species with it.

It is unfortunate that such a complete and significant specimen as ER 1470 has been made a type specimen without detailed firsthand study and comparison. Nevertheless, the name as proposed seems to fulfill the requirements of the Code and therefore to be available. No one, to our knowledge, has raised any technical objections to this proposal which might leave this name unavailable. The name has been used by Wood (1991), as *Homo rudolfensis*, in his monographic treatment of hominids from Koobi Fora.

Note that at one point Alexeev mentions a species "*Pithecanthropus heidelbergensis*" as inhabiting "Europe and North America," where from context he means North Africa, not North America. His usage of such undefined, informal terms as "Pithecanthrope", "Sinanthrope", and "Archanthrope" is reminiscent of the literature of the first half of this century.

The Origin of the Human Race

V. P. Alexeev

There is a very marked propensity in the systematics of the genus of archanthropes to exaggerate the taxonomic rank of the separate finds, as in the systematics of *Australopithecinae*. The worker singles out his find, as a rule, no matter how fragmentary it is, as a separate genus. This tendency began with Dubois who proposed the generic title which (as I have already said) is still retained by right of priority for the whole genus. Then the genus name *Sinanthropus* was proposed for Sinanthropes. Finally, generic rank was also given to Arambourg's finds in North Africa in 1954–55, and a corresponding name *Atlanthropus*. Thus, within this systematic unit, which I have taken as a genus, three genera are usually distinguished. And when we remember that the find at Mauer, near Heidelberg, in 1907, is also often given a generic status, the total number of species within the genus of archanthropes, as we see it, is increased to four.

What are the ways for understanding the true differentiation of the genus and its subdivision into species? At first glance it would seem most reasonable to follow the line of reducing the taxonomic rank of the separate fossil forms and bringing all the enumerated genera to the level of species. That way there would be four species within the genus, as I have just remarked. There are all the morphological grounds for that: the volume and structure of the brain and the structure of the skulls of *Pithecanthropus* and *Sinanthropus* have a significant similarity, though they differ in details. Similarly the structure of the teeth and jaws of Heidelberg man and the Atlanthrope are similar in spite of certain peculiar features. The geographical criterion also seems to testify in favour of the species independence of all four forms, i.e. Heidelberg man (Europe), *Atlanthropus* (North Africa), *Sinanthropus* (Eastern Asia), and *Pithecanthropus* (Southeast Asia). But a closer examination of their

From V. P. Alexeev, *The Origin of the Human Race*, translated by H. Campbell Creighton, pp. 89–94, 1986. Reprinted by permission of Progress Publishers, Moscow, Russia.

morphology forces us to introduce certain corrections to this idea.

First of all, a minimum of two of the listed types of fossil man are two successive rungs on the evolutionary ladder, viz., the Pithecanthropes and the Sinanthropes. The remarkable and widely known inquiries of Franz Weidenreich, published in 1936–45, indicate with complete certainty that *Sinanthropus* was at a higher level of evolutionary development than *Pithecanthropus*. That is indicated both by the greater volume of its brain and the higher vault of the skull, and a certain attenuation of the superciliary relief, and other less essential features of its morphology. The discovery of new remains of Pithecanthropes in recent decades fully supports the reality of these differences. That all the more strikingly contrasts Sinanthropes to Pithecanthropes since among the finds made in the '30s, *Pithecanthropus* IV (represented, it is true, only by small fragments of the skull) is distinguished by extreme coarseness and primitiveness of structure. Some workers even suggested singling *Pithecanthropus* IV out as a separate species (for which there is apparently no basis), but the find quite graphically and eloquently supports the possibility of an evolutionary contrasting of Pithecanthropes and Sinanthropes and their sequence on the evolutionary ladder. There is absolutely no doubt of the need to single out these two species within the genus *Pithecanthropus* from the evolutionary, morphological standpoint. In accordance with the priority rule they should correspondingly be called *Pithecanthropus erectus* (Dubois, 1894), i.e. upright walking Pithecanthrope, and *Pithecanthropus pekinensis* (Black, 1927), i.e. the Chinese Pithecanthrope.

As for the other two forms—Heidelberg man and Atlanthropes—many workers have repeatedly pointed out their more progressive character compared with Pithecanthropes and, on the contrary a certain similarity in that respect with Sinanthropes. The same can also be said about the more recent finds of human remains of this stage of the evolution of hominids, i.e. those from Vertesszöllös in Hungary and Bilzingsleben in the German Democratic Republic. It is difficult to speak quite definitely of a similarity between them, since all four finds represent various fragments of skulls and lower jaws. The geographical criterion is seemingly the main one in this case; one can therefore unite the populations of all four forms in one species inhabiting Europe and North America, designating it again in accordance with the priority rule *Pithecanthropus heidelbergensis* (Schoetensack, 1908).

The classification of the genus *Pithecanthropus* (now that it has been distinguished, I can use this name for all early hominids equally with the term 'archanthropes') cannot, however, be considered final, until we consider the place in the system of the so-called Javan Neanderthals, whose skulls were found at Ngandong in the valley of the River Solo in Java in 1931, and the new finds of the last two decades in Africa. In the overwhelming majority of both special studies and general works the Javan forms from the Solo River are put into the group of palaeoanthropes. But the author of the detailed morphological description of these forms, Franz Weidenreich (1951) convincingly showed that they were much closer in brain and skull structure to the Pithecanthropes and Sinanthropes than to palaeoanthropes. In brain volume they were close to Sinanthropes; many extremely primitive features were discovered in their skull structure (a very developed sagittal torus, and an unusual growth of the relief of the skull). Weidenreich's point of view was shared by Bunak (1959). Certain statistical comparisons also support it (Alexeev, 1978). But the position of these finds in the classification system is not decided merely by their similarity with members of the genus *Pithecanthropus*, noted above. Their chronological dating is very late, and they are possibly synchronous with palaeoanthropes, but might even be the earliest representatives of modern man (Oakley, 1964; Ivanova, 1965). A similarity with palaeoanthropes has been discovered in individual morphological features. The peculiar morphology of these finds, in which very primitive attributes of the genus *Pithecanthropus* are combined with certain progressive features characteristic of palaeoanthropes, and also their late chronological age make it possible to distinguish a fourth species in the genus *Pithecanthropus* and call it, employing the priority of the name, *Pithecanthropus soloensis* (Oppenoorth, 1932).

In strata at Olduvai, dated absolutely as roughly 300,000 y.a., a skull was found (designated in the literature as Olduvai II) that very much resembled that of Javan *Pithecanthropus* in the ratio of its dimensions, i.e. a very low height and great length of the brain case, a brain volume of approximately 1,000 cubic centimetres, and development of skull relief, especially of the frontal relief. The total geographical range of this form in the broad sense remains unknown, but it is clear that its typical features are repeated in the early population of Koobi Fora. This concerns two skulls whose chronological age is apparently more than 1,500,000 years (Walker, R. Leakey, 1978). One of these, designated KNM-ER 3733, had a brain volume of 850 cubic centimetres; the other, designated KNM-ER 3883, had a brain volume greater than the first, in this case around 1,000 cubic centimetres. Heberer (1963), relying on the structural features of the Olduvai II skull, and allowing for its similarity to Javan *Pithecanthropus* and the African geographical range, distinguished a new species of *Pithecanthropus*, naming it *P. leakeyi* in honour of Louis Leakey. This is a fifth species of

Pithecanthropus, in which the forms mentioned from Koobi Fora should be included, despite the fact that there is a chronological gap of more than a million years between them.

The finding of a skull, at Koobi Fora in 1972, designated KNM-ER 1470, attracted great attention. It was originally dated at 2,700,000 to 3,000,000 y.a. (Day, Leakey, Walker, Wood, 1975), but now the dating has been changed to not less than 1,600,000 y.a., but may be more than 2,500,000 years old (Walker, Leakey, 1978; M. Leakey, R. Leakey, Behrensmeyer, 1978). The brain volume was determined at first at around 800 to 820 cubic centimetres, but later, after a second, more accurate measurement, at 700 to 775 cubic centimetres. The skull is much more gracile than those of *Pithecanthropus soloensis* and *P. leakeyi*, which is expressed in both the thickness of the bones and the slight development of frontal relief. It could be thought, from that, that it is the skull of a female specimen, but the exceptional length of the face, close to the maximum for ancient and early men, forces one to think it a male skull. In that case its differences from the skulls of other Pithecanthropes are obvious. It has therefore been suggested to include a sixth species in the genus, called *Pithecanthropus rudolfensis*, from the old name of Lake Turkana. The skull KNM-ER 1813, discovered at Koobi Fora in 1973 should also apparently be included in this species; it occurred in geologically later strata dated not later than 1,200,000 y.a., but possibly belonging to an earlier time of 1,600,000 y.a. (Walker, Leakey, 1978; M. Leakey, R. Leakey, Behrensmeyer, 1978). This time we are dealing with an indisputably female individual, with a skull of very small dimensions and a brain volume of not more than 500 cubic centimetres. In its structural features, however, it is similar to KNM-ER 1470. Such dimensions resemble Central African pigmies; perhaps there was a tendency to dwarfishness in Africa from the earliest stages of the history of the hominid family, which stems from many causes (environmental influences, shifts in growth processes, appropriate selection). We await further discoveries.

The place of *Pithecanthropus* in the history of hominids is determined by the chronological range of the finds. For the Javan *Pithecanthropus* we have a date of 1,000,000 to 1,500,000 y.a.; the African Pithecanthropes may even be older. The Solo hominids, as already indicated, are most probably very late. *Pithecanthropus* thus represents, on the whole, the first group of men proper, which existed for a very long time; arising from *Australopithecus*, it was an intermediate link between the subfamily of *Australopithecinae* and the genus *Homo*. Some forms of *Pithecanthropus* continued to exist, obviously, simultaneously with *Homo*.

References

The following references are taken from the List of Recommended Reading that accompanied Alexeev's entire work, *The Origin of the Human Race*. Some of the authors and works cited in this excerpt were not included in that list.

Alexeev, V. P. *Paleoantropologia zemnogo shara i formirovanie chelovecheskikh ras* (The Palaeoanthropology of the World and the Forming of the Human Races), Nauka, Moscow, 1978.

Day, M., Leakey, R., Walker, A., Wood, B. New hominids from East Rudolf, Kenya (I). *Am. J. Phys. Anthrop.* (new series), 1975, *42*: 461–476.

Dubois, E. *Pithecanthropus erectus, eine menschenähnliche Ubergangsform von Java* (Batavia, 1894).

Leakey, M. G., Leakey, R. E. (Eds.) *The Koobi Fora Research Project*, Vol. 1. *The Fossil Hominids and an Introduction to Their Context 1968–1974* (Clarendon Press, Oxford, 1978).

Oakley, Kenneth. *Frameworks for Dating Fossil Man* (Aldine, Chicago, 1964).

Walker, A., Leakey, R. E. The hominids of East Turkana. *Scientific American*, 1978, *239*, 2.

Weidenreich, F. The duration of life of fossil man in China and the pathological lesions found in his skeleton. *Chinese Medical Journal*, 1939, *55*.

Weidenreich, F. The skull of *Sinanthropus pekinensis*: a comparative study of primitive hominid skulls. *Pal. Sinica* 1943 (whole series, No. 127), new series D, No. 10.

Part II

Influential Interpretations of Hominid Taxonomy

Part II

16

The Cold Spring Harbor symposium volume of 1950 contains several historically important papers for paleoanthropology, including that of Ernst Mayr reprinted here. This volume appeared during the period when the "modern synthesis" of evolutionary theory was becoming well-established through the combined publications of Mayr, G. G. Simpson, T. Dobzhansky, and others. The Cold Spring Harbor symposium took place during a significant period of contact between paleoanthropologists and evolutionary biologists. As mentioned above, the modern period of fossil hominid studies began to develop after World War II as the number of trained paleoanthropologists increased.

Zoologist Ernst Mayr is known for his systematic studies of birds and for his interests in evolutionary theory and history. In this influential paper he dispenses with almost all the fossil taxa and names in common use earlier in this century, reducing the hominid family to a single genus, *Homo*, with only three successive species. Most of the variation apparent in the human fossil record is attributed by Mayr to individual, population, or sub-specific causes. Note that he is tentative, and certainly not dogmatic, in his presentation of these suggestions. It is now clear that he goes too far towards a "lumping" extreme in his reaction against the ridiculously over-split hominid taxonomy of the day. However, his example did establish a minimum taxonomy to which other schemes could be compared. Tattersall (selection 19) comments on the importance of this paper for later debates about the number of extinct hominid species.

Note that Mayr's paper was published several years before the exposure of the Piltdown fraud, and therefore that he mentions this specimen as a legitimate fossil.

Taxonomic Categories in Fossil Hominids

Ernst Mayr

It is one of the most fruitful procedures of modern science to bring specialists of various fields together to discuss the problems that concern the zone of overlap of their fields. Not possessing any first-hand knowledge of paleoanthropology, my own contribution to the question of the taxonomic categories of fossil man will be that of a systematist. Significant progress has been made within recent years among biologically thinking taxonomists in the understanding of the categories of subspecies, species, and genus, and it is my hope that this knowledge may help in a better understanding of fossil man.

The whole problem of the origin of man depends, to a considerable extent, on the proper definition and evaluation of taxonomic categories. But, there is less agreement on the meaning of the categories species and genus in regard to man and the primates than perhaps in any other group of animals. Some anthropologists, in fact, imply that they use specific and generic names merely as labels for specimens without giving them any biological meaning. The late Weidenreich, for example, stated that in anthropology "it always was and still is the custom to give generic and specific names to each new type without much concern for the kind of relationship to other types formerly known." Broom (1950) likewise states, "I think it will be much more convenient to split the different varieties [of South African fossil ape-man] into different genera and species than to lump them." The result of such standards is a simply bewildering diversity of names. In addition to various so-called species of *Homo*, the following names for various hominid remains have been found by me in the literature: *Australopithecus*,

From Ernst Mayr, "Taxonomic Categories in Fossil Hominids," *Cold Spring Harbor Symposia on Quantitative Biology*, Vol. 15, pp. 109–118 (1950). Reprinted by permission of CSHSQB Publications, Cold Spring Harbor, NY.

Plesianthropus, Paranthropus, Eoanthropus, Giganthopithecus, Meganthropus, Pithecanthropus, Sinanthropus, Africanthropus, Javanthropus, Paleoanthropus, Europanthropus, and several others. No two authors agree either in nomenclature or in interpretation. It seems to me that an effort should be made to give the categories species and genus a new meaning in the field of anthropology, namely, the same one which in recent years has become the standard in other branches of zoology.

A re-evaluation of the terminology of hominid taxonomy is facilitated by the fact that in recent years a magnificent body of new data has been accumulated by anthropologists, partly based on comparative anatomical studies and partly on significant new discoveries of fossil man in southeast Asia and in eastern and southern Africa.

The nomenclatorial difficulties of the anthropologists are chiefly due to two facts. The first one is a very intense occupation with only a very small fraction of the animal kingdom which has resulted in the development of standards that differ greatly from those applied in other fields of zoology, and secondly, the attempt to express every difference of morphology, even the slightest one, by a different name and to do this with the limited number of taxonomic categories that are available. This difference in standards becomes very apparent if we, for example, compare the classification of the hominids with that of the *Drosophila* flies. There are now about 600 species of *Drosophila* known, all included in a single genus. If individuals of these species were enlarged to the size of man or of a gorilla, it would be apparent even to a lay person that they are probably more different from each other than are the various primates and certainly more than the species of the suborder Anthropoidea. What in the case of *Drosophila* is a genus has almost the rank of an order or, at least, suborder in the primates. The discrepancy is equally great at lower categories, as we shall presently see. It is not mere formalism to try to harmonize the categories of anthropology with those of the rest of zoology. Rather, the evaluation of human evolution depends to a considerable extent on the proper determination of the categories of fossil man.

There are two recent developments in general systematics that will be particularly helpful in our efforts. The first one is that the biological meaning of the categories species and genus is now better understood than formerly, and second, that, in the attempt to close the gap between the complexity of nature and the simplicity of categories, the number of existing categories has been augmented by intermediate and group categories, such as "local population" or "local races" and "subspecies groups." The adoption of these intermediate categories facilitates classification without encumbering nomenclature.

The Taxonomic Categories

The work in the new systematics has led to a far-reaching agreement among zoologists on the meaning of the categories subspecies, species, genus, and family. In the following an attempt shall be made to see how far the current usage of these categories can be extended to fossil hominids and what such a reclassification means in terms of human evolution.

The genus: The genus is a taxonomic category for a group of related species. It is usually based on a taxonomic group that can be objectively defined. However, the delimitation of these groups against each other, as well as their ranking, is frequently subjective and arbitrary. A conventional definition of the genus would read about as follows: "A genus consists of one species, or a group of species of common ancestry, which differ in a pronounced manner from other groups of species and are separated from them by a decided morphological gap."

Recent studies indicate that the genus is not merely a morphological concept but that it has a very distinct biological meaning. Species that are united in a given genus occupy an ecological situation which is different from that occupied by the species of another genus, or, to use the terminology of Sewall Wright, they occupy a different adaptive plateau. It is part of the task of the taxonomist to determine the adaptive zones occupied by the various genera. The adaptive plateau of the genus is based on a more fundamental difference in ecology than that between the ecological niches of species.

Unfortunately, there is no such thing as a recognized or absolute generic character. This was known already to the earlier taxonomists, in fact, Linnaeus stated, "It is the genus that gives the characters, and not the characters that make the genus." The genus is a group category and it defeats the object of binomial nomenclature to place each species into a separate genus, as has been the tendency among students of primates.

The acceptance of the new concept of biologically defined polytypic species (see below) necessitates the upward revision of all other categories (Mayr, 1942). Often what was formerly a group of allopatric species is now a single polytypic species with numerous subspecies. To leave each of these polytypic species in a separate genus deprives the genus of its significance as a truly collective category. I shall illustrate this need for the combining of genera by an example. Gorilla and chimpanzee are two excellent species which, as Professor Schultz has shown, differ from each other by a wealth of characters. At one time several species of gorillas and of chimpanzees were recognized, but the allopatric forms within the two species are now considered subspecies. Being left with one species of gorilla and one species of chimpanzees, we are confronted by the question whether or not they are sufficiently different to justify placing them in different genera. A specialist of anthropoids impressed by the many differences between these species may want to do so. Other zoologists will conclude that the differences between the two species are not indicative of a generic level of difference when measured in the standards customary in most branches of zoology. To place these two anthropoids into two separate genera defeats the function of generic nomenclature and conceals the close relationship of gorilla and chimpanzee as compared with the much more different orang and the gibbons. Recognizing a separate genus for the gorilla would necessitate raising the orang and the gibbon to subfamily or family rank as has indeed been done or suggested. This only worsens the inequality of the higher categories among the primates.

The same is true for the fossil hominids. After due consideration of the many differences between Modern man, Java man, and the South African ape-man, I did not find any morphological characters that would necessitate separating them into several genera. Not even *Australopithecus* has unequivocal claims for separation. This form appears to possess what might be considered the principal generic character of *Homo*, namely, upright posture with its shift to a terrestrial mode of living and the freeing of the anterior extremity for new functions which, in turn, have stimulated brain evolution. Within this type there has been phyletic speciation resulting in *Homo sapiens*.

The claim that the many described genera of hominids and Australopithecines have no validity, if the same yardstick is applied that is customary in systematic zoology, is based on two major points. Both of these are admittedly somewhat vulnerable. One is the overall picture of morphological resemblance with a deliberate minimizing of the brain as a decisive taxonomic character. To this point we shall return presently. The other point is the assumption that all these forms, including *Australopithecus*, are essentially members of a single line of descent. Additional finds might easily disprove this. However, taking all the available evidence together, it seems far more logical and consistent at the present time to unite the hominids into a single genus than to continue the current multiplicity of names.

This re-evaluation of the generic status of the fossil hominids forces us to consider also the categories above the genus. Does *Homo* belong to a separate family Hominidae? The morphological differences between *Pongo*, the genus to which the chimpanzee and gorilla belong, and *Homo* are so slight that there seems to be no justification for placing them in separate families. There is even less justification for placing South African man in a separate subfamily, the Australopithecinae. The most primitive known hominids, those of South Africa, combine certain typical hominid characters, such as upright posture, with others that are usually considered simian, such as small size of brain and protruding face. It is noteworthy, however, as pointed out by several investigators, that these hominids, even at this primitive stage, lack certain other simian features that were formerly considered as primitive: powerful canines, large incisors, a sectorial form of the first lower premolar, an exaggerated development of the supra-orbitals, a simian shelf, and powerful brachiating arms. It now appears probable that many of these characters are functional specializations which were acquired by the anthropoid apes after the hominid line had branched off.

The fact that the hominids lack these specializations has been used by some authors as evidence to postulate a very early human origin and a very isolated position of the hominid branch. This is by no means the only possible interpretation. Rather it seems to me that most of these typical characters of the living anthropoids may well be a single character complex evolved in response to a highly arboreal mode of living. It now appears probable that the African anthropoids, the orang, and the gibbons, may have acquired most of these characters independently and are therefore, in a sense, a polyphyletic group. The available evidence seems to indicate to me that man may be more closely related to the gorilla-chimpanzee group than this group is either to the orang or to the gibbons. The degree of similarity in certain morphological traits cannot necessarily be used to measure degree of phylogenetic relationship. The arboreal, brachiating large anthropoids are exposed to a similar type of selection and will therefore evolve in a parallel, if not convergent, manner. When the *Homo*-line acquired upright posture it entered a completely different adaptive zone and became exposed to a severely increased selection pressure. This must have resulted in a sharp acceleration of evolutionary change leading to the well-known differences between man and the living anthropoids. This factor must be taken into consideration when the phylogeny of man and the anthropoids is reconstructed. It would therefore appear to be misleading from the purely morphological-phylogenetic point of view to separate man from the anthropoid apes as a special family.

It would be equally misleading to go to the other extreme and to use the evidence of the somewhat independent evolution of man and the various anthropoids as a means to deny their close relationship.

Denying the genus *Homo* family rank is based on purely morphological considerations. It does not take into account man's unique position in nature. Man has undoubtedly found an adaptive plateau that is strikingly different from that of any other animal. There are some who feel that there is only one way by which to emphasize this uniqueness of man, namely, by placing *Homo* into a separate family. The conventional standards of taxonomy are insufficient to decide what is correct in this case.

From the purely biological point of view man is certainly at least as different as a very good genus. We have thus the evolution of a new higher category in the geologically short period of one to two million years. This is another significant illustration for the rapidity by which one major taxonomic entity can be transformed into another one, without any jumps.

The subspecies: Before we can attempt to answer the question how many species of fossil man have existed, we must say a few words on infraspecific categories. The species of the modern systematist is polytypic and multidimensional. It has the geographical dimensions of longitude and latitude and also the time dimension. It is polytypic because it is composed of lower units, such as subspecies and local populations. Customarily in anthropology, distinct local populations have been referred to as races, and a similar custom exists in some branches of zoology as, for example, in ichthyology.

The amount of geographical variation and the degree of difference among the geographical subdivisions of a species are different from case to case. Some species appear quite uniform throughout their entire range; other species have a few or many more or less well defined subspecies. For instance, the two African forest anthropoids, chimpanzee and gorilla, show only a moderate amount of geographical variation, although both have well-defined subspecies, and attempts have been made to split the chimpanzee into several species. Geographical variation is much more pronounced in the orang and even more so in some of the South American monkeys where geographical races are often different enough to be considered full species by conservative authors.

Modern man is comparatively homogeneous because there is much interbreeding between different tribes and races. Still, we find in close neighborhood to each other such strikingly different races as bushmen and Bantus in South Africa, or the Congo pygmies and

Watusi in central Africa, or the Wedas and Singhalese in Ceylon. There is much indirect evidence that primitive man was much more broken up into small scattered tribes with little contact with each other, intensely subject to local selective factors.

In addition to this much greater geographical variation of primitive man, there is evidence also of greater individual variation (including sexual dimorphism). The variability of Mt. Carmel man has been commented upon in the literature. It seems possible, if not probable, that the various South African finds, *Australopithecus*, *Plesianthropus*, and *Paranthropus*, might well be age or sex stages of a few related tribes, notwithstanding Broom's (1950) assertions to the contrary.

Differences between young and adult and between male and female appear to be greater in the gorilla and orangutan than they are in modern man. Variability may increase or decrease in the course of evolution. Abundant proof for this statement can be found in the paleontological literature. I interpret the available literature to indicate that primitive man showed more geographical as well as individual variation than modern man.

Why primitive man should have been more variable than modern man is not entirely clear. A study of the family structure of anthropoids might shed some light on this problem. Perhaps there was a greater functional difference between male and female than in modern man. Perhaps the ancestral hominids had a system of polygamy that would favor the selection of secondary sex characters in the male. We don't know. Whatever the reasons, we should not use the variability within populations of modern man as a yardstick by which to judge the probable variability of extinct populations.

This point is important because it bears on the question whether or not more than one species of hominid has ever existed on the earth at any one time. Indeed, all the now available evidence can be interpreted as indicating that, in spite of much geographical variation, never more than one species of man existed on the earth at any one time. We shall come back to this point later.

The species: As described in several publications, the concept of the species has undergone a considerable change during recent years. The morphological and typological species of the early taxonomists has been replaced by a biological species. The species is now defined "as a group of actually or potentially interbreeding natural populations that is reproductively isolated from other such groups." When this concept is applied to man, it is at once obvious that all living populations of man are part of a single species. Not only are they connected everywhere by intermediate populations but even where strikingly distinct human populations have come in contact, such as Europeans and Hottentots, or as Europeans and Australian aborigines, there has been no sign of biological isolating mechanisms, only social ones.

The problem of species delimitation is much more difficult with respect to fossil man. How shall we determine which populations are "actually or potentially interbreeding"? It is evident that we must use all sorts of indirect clues. The first concrete problem is what types of fossil man should be included in the species *Homo sapiens*. Cro-Magnon man is so nearly identical with *Homo sapiens* that its inclusion in that species is not doubted by any serious student.

The problem of Neanderthal man is much more difficult. Should he be included in the same species as modern man or not? When the first finds of Neanderthal man were made there seemed to be no problem. These fossils were characterized by distinct morphological features and were clearly replaced by modern man in Europe on a distinct chronological level. There is no morphological or cultural intermediacy. Additional finds, however, have caused various difficulties. In Palestine the Mt. Carmel finds belong to a population that combines some features of Neanderthal with some of modern man. It is immaterial whether we interpret this as a hybrid population, as an intermediate population, or as a population ancestral to both. The fact remains that Mt. Carmel man makes the delimitation of modern man from Neanderthal exceedingly difficult, if not impossible, as pointed out by Dobzhansky (1944). Weidenreich supported the theory that modern man was a direct descendant of Neanderthal man. Boule and others have raised serious objections to this theory. But how can we reconcile the apparently incompatible views that modern man and Neanderthal are conspecific and that modern man is *not* a descendant of typical European Neanderthal? A possible clue is furnished by the hominids that were widespread in Europe in mid-Pleistocene. The skulls of Steinheim, Swanscombe, and of Fontéchevade combine features of modern man and of Neanderthal man, together with primitive and specialized features of their own. They lived apparently in inter-glacials and were more closely linked with a warm climate than Neanderthal man.

If I understood the evidence correctly, it is possible to interpret these early European fossils as remains of populations of *Homo* that were ancestral both to *sapiens* and to "classical" Neanderthal and from which these two forms evolved by geographical variation. Tentatively the working hypothesis can be made that Neanderthal in its classical form was a geographical race that occurred in central Europe and was represented in Africa by Rhodesian man and in Java by Njandong man, while a more *sapiens*-like population occurred at the same period as some of these Neanderthaloids either in north Africa or western Asia or in some other area that has not yet yielded remains of fossil man. When *sapiens* began to expand and spread, he eliminated the other contemporary races just as the white man drove out the Australian aborigines and the North American Indians. The process of elimination of the Neanderthal characters in mixed populations was presumably helped by selection preference in favor of the characters of modern man.

It is very probable that additional finds will make the delimitation of *sapiens* against Neanderthal even more difficult. It seems best to follow Dobzhansky's suggestion and to consider the two forms, as well as the ancestral group that seems to combine their characters, as a single species.

Homo erectus: Java and Peking man are sufficiently distinct from modern man so that they have to be considered a separate species, which must be called *Homo erectus.* This is true regardless of the fact that on Java, at least, Njandong and Wadjak man may have formed a practically unbroken chain of hominids leading from Java man to modern man. Peking man (*Homo erectus pekinensis*) is, on the whole, so similar to Java man that it should be considered merely subspecifically distinct, as I proposed previously (Mayr, 1944).

In spite of its obvious similarity to *Australopithecus* too little is known of the still earlier Java *Meganthropus* to assure a correct classification. This is even more true of *Giganthopithecus* whom some authors consider hominid and others anthropoid. One thing about *Giganthopithecus* is, however, very probable, namely, that it was not necessarily a giant in spite of its giant teeth. Jaws and teeth of early fossil man were relatively much larger than they are in modern man.

Homo transvaalensis: South African ape-man again is one level further back and is sufficiently far removed from Java man to be considered a full species. Actually, no less than three genera and five species of South African ape-man were described which, in Broom's terminology, have the following names: *Australopithecus africanus* 1925 (Taungs), *A. prometheus* 1947 (Makapan), *Plesianthropus transvaalensis* 1936 (Sterkfontein), *Paranthropus robustus* 1938 (Kromdraai), and *Paranthropus crassidens* 1949 (Swartkrans). Most of these names may not have any validity, according to the Rules of Zoological Nomenclature, Article 25A, as revised in 1930. According to these Rules a name has validity only if the description includes diagnostic characters. Since one of these names was based on a child, another on an adult female, a third on an adult male, an enumeration of diagnostic differences is virtually impossible. The extant skulls are somewhat altered in shape due to crushing, and the fact that the cephalic index in the Taungs child is 62.4 while it is 83.5 in the Sterkfontein male is therefore not as significant as Broom thinks. Nor is the fact that the finds are associated with different faunas. Contemporary modern man can be found associated with okapis or elephants or tigers or kangaroos, or South American edentates or with polar bears. The various finds of South African man are presumably not contemporary, but there is nothing in the evidence that has so far been presented (e.g. Broom, 1950) that would prove that more than one species is involved.

Until a real taxonomic distinction has been established, it will be safer and more scientific to refer to the different South African fossils by vernacular names. There is no danger of confusion if we speak of the Sterkfontein or Makapan finds, while it implies an obviously erroneous conclusion, namely that of generic distinctness, if we refer to them as *Plesianthropus* and *Australopithecus*. New discoveries are still being made in these cave deposits and many of those that have already been made have not yet been fully worked out. There is good reason to believe that it will be firmly established in the not-too-distant future how many different tribes, temporal subspecies, or even species of South African ape-man once existed. To consider them all as one species is the simplest solution that is consistent with the available evidence.

A more important question is whether South African man is ancestral to modern man or merely a specialized or aberrant sideline. The exact dating of these fossils has not yet been achieved but they are believed to be very early Pleistocene or latest Pliocene, in fact, they presumably ranged over a considerable period of time. There is thus no definite chronological reason why the South African ape-man could not be considered a possible ancestor of modern man. The principal objection that has been raised is that South African man shows a combination of characters that "should not" occur in an early hominid. This argument is based on typological considerations. Adherents of this concept believe that missing links should be about half-way between the forms they connect and that they should be half-way in every respect. This undoubtedly is not the case with *Australopithecus*. It is apparently amazingly like modern man in its upright posture, structure of the pelvis, and other features, while it is very simian in its massive mandibles, large molars, prognathism, and small brain. *Australopithecus* lacks those specializations that stamp gorilla, orang, and gibbon as typical anthropoids.

The peculiar combination of characters that is found in *Australopithecus* is due to the fact that during evolution of man different characters evolved at different rates. If we would set the point where the human line branched off from the other anthropoids as zero and the *Homo sapiens* stage as 100, we might give arbitrarily the following points to the various organs of *Australopithecus*: pelvis, 90; premolars, 75; occipital condyles, 80; incisors, 55; the setting of the brain case, 70; shape of the tooth row, 70; the profile of the jaw, 30; the molar teeth, 40; the brain, 35; etc. It is obvious that one type does not change into another type evenly and harmoniously, but that some features run way ahead of the others.

The inability to understand this has been the reason for Weidenreich's insistence that *Eoanthropus* was an artifact. He maintained with respect to Piltdown man: "Form and individual features of the brain case are generally acknowledged as those of modern man; those of the lower jaw have anthropoid characteristics. Therefore, both skeletal elements cannot belong to the same skull." As a matter of fact, the skull cap is not strictly modern nor is the jaw strictly anthropoid and the recent fluorine content determination by Oakley (1950) indicates that, indeed, jaw and skull cap may be of the same geological age.

It may take a long time before the Piltdown puzzle is completely cleared up. As a working hypothesis it might be suggested that Piltdown man represents a geographical race of man that was restricted to northwestern Europe. Some of the characters, particularly in the jaw, appear to be specializations rather than indications of primitiveness, perhaps developed in connection with the large size of the individual. The phylogenetic and chronological relationship of Piltdown to the other hominid finds indicated by the words Heidelberg, Steinheim, Swanscombe, and Fontéchevade still remains to be determined.

The simplified nomenclature of fossil man: Reducing the bewildering assortment of genera and species of hominids to one genus with three species results not only in simplicity but it also makes certain conclusions obvious that were previously not apparent. Before discussing these conclusions, however, I might point out some of the disadvantages of such a simplified classification.

There have been two trends in human evolution as, indeed, there are in the evolution of all organisms. First of all, there is a continuous evolutionary change in time, the so-called phyletic evolution, starting in the hominids with the most simian forms and ending with modern man. Simultaneously a centrifugal force has been operating, namely, geographical and other local variation, which tries to break up the uniform human species. This geographic variation leads to the formation of races and subspecies, and if this trend would go to completion, to the formation of new separate species. There are all sorts of intermediate stages in both these trends and it is obvious that all the many possible differences and gradations between the various kinds of hominids cannot be expressed completely in the simple nomenclature of species, genus, and subspecies.

For instance, man as he exists today, has pronounced racial groups, such as the Whites, Negroes, and Mongoloids, which might well deserve subspecific recognition. But there are minor racial differences within each of these subspecies. Furthermore, preceding modern man there have been types of *Homo sapiens* that are now extinct, like Cro-Magnon man and his contemporaries. This, no doubt, is a different level of subspecies from those of living man. Neanderthal man is a third level, and the pre-Neanderthal man, who combines certain features of *sapiens* and Neanderthal, is a fourth level. It is unsatisfactory for biological, as well as for practical reasons, to treat each of these levels as a separate species. On the other hand, combining them into a single species conceals the pronounced differences between these levels and reduces the taxonomic difference between Neanderthal and modern man to the level of difference between White Man and Negro. How can this be avoided?

First of all, we must realize that no system of classification and nomenclature can ever hope to express adequately the complicated relationships of natural populations. However, by giving species and genus the well-defined meanings that we have assigned to these categories, we make at least an attempt to standardize taxonomic categories and make them comparable. A possible solution to our particular difficulties may come from a refinement of the levels of infraspecific categories. In addition to the subspecies we may use such infrasubspecific categories as "race" and "local population," as well as the suprasubspecific category of the "subspecies group." Hence, we should be guided by the following practical rules:

1. Not to assign a formal name to any local population or race that does not deserve subspecific rank.

2. To give trinomials to all forms that do not deserve higher than subspecies rank.

3. To group together as subspecies groups all those subspecies within a species that form either geographical or chronological groups.

    Such subspecies groups in *Homo sapiens*, for instance, might be:

    (a) modern man

    (b) Neanderthal group

    (c) pre-Neanderthal group

4. Not to give formal generic and specific names to new fossil finds that are not sufficiently known. Vernaculars, such as "Steinheim man" or "Piltdown man," are just as useful and much less misleading. The formal application of generic and specific names simulates a precision that often does not exist. To give the impression of an unjustified precision is as much of a methodological error as to make calculations to the fifth decimal when the accuracy of the original data extends only to the first decimal.

Anthropologists should never lose sight of the fact that taxonomic categories are based on populations, not on individuals. Different names should never be given to individuals that are presumably members of a single variable population.

Conclusions

The arranging of all finds of fossil hominids into a single genus with three species helps to focus attention on the following conclusions.

The question of the "missing link." Ever since there has been an appreciation of man's anthropoid origin there has been a search for the "missing link." Some anthropologists may disclaim this and say that they realize the gradual evolution of mankind but the fact remains that accurate criteria of humanhood are elaborated even in the most recent literature, such as Sir Arthur Keith's criterion of the brain volume of 750 cc.

The analysis of this problem will be facilitated by the realization that it is an oversimplification to use in this case the uninomial alternative "ape" versus "man." Taxonomists know by experience the inadequacy of uninomialism. Classifying man binomially as *Homo sapiens*, it at once becomes apparent that we must look for two missing links, namely that which connects *sapiens* with his ancestor and that which connects *Homo* with his ancestor, Or, to express this differently, the two points of interest are the one on the phyletic line of man where he reached the *sapiens* level and second the place where the *Homo* line branched off from the other primates.

Let us look more closely at these two problems of the origin of man. The branching off of *Homo* from the other anthropoids was a case of orthodox speciation distinguished only by the fact that the new species simultaneously reached a new adaptive plateau. It is now evident, as has been stated by many authors, that a change in the mode of locomotion and a corresponding alteration of the entire organization of the body, in other words, the assuming of the upright posture, were the essential steps that led to the evolution of *Homo*. This evolutionary trend apparently affected first the pelvis and posterior extremities followed closely by the anterior extremities. The corresponding re-organization of the skull lagged apparently behind. It is therefore singularly difficult to localize both in time and space this important evolutionary step of the attainment of the upright posture with the help of jaw and tooth fragments, such as constitute most of the primate and anthropoid remains in eastern Africa during Pliocene and Miocene.

To determine the exact point in the phyletic evolution of *Homo* where the *sapiens* level was reached, is quite impossible. It was a very gradual process leading from *erectus* to *sapiens* and no particular form can be singled out as the missing link. However, there is a lower level in the phyletic evolution of *Homo* that is of special evolutionary interest, namely, the level at which the hominids first displayed those intellectual qualities that are considered distinctly human rather than simian.

Attempts have been made to measure the attainment of this *Homo* level in terms of brain size. This method is fraught with difficulty. First of all, brain size is to some extent correlated with body size. If, for instance, a large gorilla should have a brain of 650 cc. this is not at all necessarily equivalent to the brain of a fossil hominid of 650 cc., if that hominid were much smaller than a gorilla. If the brain of the gorilla averages one-fourth larger than that of the chimpanzee, it does not mean that he is on the average 25 per cent more intelligent. The correlation between brain size and intelligence is very loose. There is good evidence that the brain size of late Pleistocene man may have averaged larger than that of modern man. If true, this does not mean necessarily that there has been a deterioration of man's intelligence since the Pleistocene, for intelligence is determined not only by brain size. It is, of course, still unknown what neurological structures affect intelligence but the folding of the cortex and all sorts of specializations within the cortex appear to be as important as size. It is therefore dangerous, in fact, outright misleading, to use size as an absolute criterion and to say that the *Homo* stage was reached when brain size reached a level of 700 or 750 cc.

It has been suggested to measure the attainment of the human level by some cultural achievement, such as the use of fire, rather than by an anatomical standard like brain size. This is unquestionably a superior approach, but has the practical difficulty that the first moment of fire making was not fossilized and can never be dated accurately. However, the first making of fire may have occurred not much after the first use of tools by hominids and some lucky finds may shed light someday on the period when that occurred. South African man was presumably already a user of tools, and the first use of tools may be coincident with the evolution of South African man.

Speciation in man: In the strict sense of the word, speciation means the origin of discontinuities through the origin of reproductive isolating mechanisms. How often has man speciated? The answer is that he has speciated only once if our assumption is correct that never more than one species of man existed on the earth at any one time. This single event of speciation was the branching off of *Homo* from the anthropoid stock. That some fairly distinct hominid remains have been found in approximately contemporary deposits does not prove their specific distinctness. The sub-division of the human species into independent tribes favors diversification. If fossils of Congo pygmies and of Watusi were to be found in the same deposit by a paleontologist, a million years hence, he might well think that they belonged to two different species. As stated previously, the known diversity of fossil man can be interpreted as being the result of geographic variation within a single species of *Homo*. This led to the evolution of such aberrant types as Piltdown man of England, but apparently nowhere to the simultaneous occurrence of several species of *Homo*. What is the cause for this puzzling trait of the hominid stock to stop speciating in spite of its eminent evolutionary success? It seems to me that the reason is man's great ecological diversity. Man has, so to speak, specialized in despecialization. Man occupies more different ecological niches than any known animal. If the single species man occupies successfully all the niches that are open for a *Homo*-like creature, it is obvious that he cannot speciate. This conforms strictly to Gause's Rule. Also man is apparently slow in establishing isolating mechanisms. This is indicated by the numerous instances of incomplete speciation in the history of the hominids. In no case was this speciation completed because the segregating populations were either absorbed by intermarriage or exterminated. Man is apparently particularly intolerant of competitors. The wiping out or absorption of primitive populations by culturally more advanced or otherwise more aggressive invaders, which we have witnessed so many times during the eighteenth and nineteenth centuries in Australia, North America, and other places, has presumably happened many times before in the history of the earth. The elimination of Neanderthal man by the invading Cro-Magnon man is merely one example.

There is one striking difference between man and most of the animals. In animals whenever there is competition between two subspecies the one that is better adapted for a specific locality seems to win out. Man, who has reached such a high degree of independence from the environment, is less dependent on local adaptation, and a subspecies of man can quickly spread into many geographically distant areas if it acquires generalized adaptive improvements such as are described by the social anthropologist. Such improvements do not need to and probably often do not have genetic basis. The authors who have claimed that man is unique in his evolutionary pattern are undoubtedly right. Even though the phyletic evolution of man will continue to go on, the structure of the human species at the present time is such that there appears to be very little chance for speciation, that is, for the division of the single human species into several separate species.

Summary

1. There is no conclusive evidence that more than one species of hominids has ever existed at a given time.
2. It is proposed to classify fossil and recent hominids tentatively into a single genus (*Homo*) with three species (*transvaalensis, erectus, sapiens*).
3. The recognition of subspecies groups within the species facilitates classification.
4. The ecological versatility of man and his slowness in acquiring reproductive isolating mechanisms have prevented the breaking up of *Homo* into several species.

References

Broom, R., 1950, The genera and species of the South African fossil ape-man. *Amer. J. Phys. Anthrop.* (N.S.) *8*:1–13.

Dobzhansky, Th., 1944, On species and races of living and fossil man. *Amer. J. Phys. Anthrop.* (N.S.) *2*:251–265.

Mayr, E., 1942, *Systematics and the Origin of Species.* New York: Columbia University Press.

_____, 1944, On the concepts and terminology of vertical subspecies and species. *Natl. Res. Council Bull. 2*:11–16.

Oakley, K. P., 1950, New evidence on the antiquity of Piltdown man. *Nature*, Lond. *165*:379–382.

Straus, W. L., Jr., 1949, The riddle of man's ancestry. *Quart. Rev. Biol. 24*:200–223.

Discussion

LASKER: I wonder how Professor Mayr would justify considering men, gorilla and orang to belong to three different genera, but man, the australopithecines and Piltdown man to belong to one genus. Considered purely morphologically, an animal with "ape-like" jaw and "man-like" skull cap, such as Piltdown, if the parts really belong to one form, differs greatly from one with anthropoid brain case and more human jaw, such as *Australopithecus*.

Incidentally, I can think of no adequate anatomical reason for associating the Heidelberg jaw with "*Pithecanthropus*" rather than with the Neanderthaloids.

MAYR: Since there are no absolute generic characters (see above), it is impossible to define and delimit genera on a purely morphological basis.

MONTAGU: I think it should be noted that convenience can be pushed to such a point that it becomes a confounded nuisance. When, to paraphrase A. E. Housman, the ambiguity of language is brought in to add to the already existing confusion of thought, as has been done in the anthropological taxonomy of fossil man, confusion is rendered worse confounded, and "convenience" becomes an impediment to clear thinking and further progress. The terminology of palaeoanthropology provides an unfortunate example of the systematics of confusion. When types such as Java man and Peking man can be referred to by generic names—*Pithecanthropus* and *Sinanthropus*—when, in fact, they represent no more than two subspecies or geographic races, it were high time that we did something to bring the taxonomic practice of palaeoanthropology more into line with its own theory and the practice of the newer systematists. I should therefore seriously suggest that a committee, consisting of such men as Drs. Mayr and Dobzhansky, and several palaeoanthropologists, be appointed to consider the matter of revising the nomenclature which is at present confusing the field of palaeoanthropology.

SCHULTZ: It has been long overdue and is most welcome that expert taxonomists come to the aid of anthropologists in modernizing the nomenclature of the Hominidae which has reached such absurdities as the *generic* separation of Java man and Peking man. Doctor Mayr's highly valued proposals for taxonomic reforms are rather startling to anthropologists who still hope that their Latin or Greek binominal terms can imply consistently degrees of distinction. To recognize only three species of *Homo*, of which one is assigned to the Australopithecines and another to everybody from Neanderthal man to ourselves, is a very sudden jump from one, old extreme to an opposite, new extreme, even though the latter is undoubtedly more consistent with general usage in modern systematics.

Doctor Mayr's interesting suggestion that gorilla and chimpanzee represent only different species of one genus, because the skulls of the latter are very similar to female skulls of the former, deserves comment as an excellent example to remind us, how unjustifiable it can be at times to assume close relationship between animals on the basis of close cranial resemblance alone (in one sex only). Even though the skulls of adult female gorillas differ from those of adult chimpanzees to only limited degrees, the skulls of adult male gorillas are vastly different, because sex differences in gorillas are extremely pronounced in contrast to those of chimpanzees which are generally less marked than in man. In other bodily features the two African apes can differ enormously. For instance, the proportionate size of the testes is extremely small in gorillas, whereas extremely large in chimpanzees, the female sexual skin shows very different changes in the two types, chimpanzees have relatively huge outer ears, gorillas very small ones, the hand is long and slender in chimpanzees, but relatively broader in gorillas than in any other simian primates, etc. If these and many other constant distinctions are mere species differences, then the comparatively few known differences between Peking man and Rhodesian man, e.g., would deserve at most subspecific rank.

With these remarks I wanted to point out chiefly that it will require more such profitable meetings between anthropologists, primatologists and taxonomists before we can expect a generally acceptable, detailed, and much needed change in the systematics of past and present higher primates.

WASHBURN: I feel that it is useful to have separate names to specify each of the major adaptive groups of the primates. Traditionally, such groups as galagoes, lorises, or monkeys were given family status. If this continues to be done, then it is convenient to put apes and men in separate families, indicating the difference between arboreal brachiation and bipedal ground living. Within the human family two groups may be recognized, the small-brained man apes and large-brained, tool using man. It may be well to keep the generic names *Australopithecus* and *Homo* for these. I thoroughly agree with Dr. Mayr on the desirability of abolishing most of the genera he has mentioned. But too few names can be confusing, and I wonder if Dr. Mayr would feel this suggestion is counter to his idea.

MAYR: An unequivocal decision on the ranking of the groups that are included in the higher categories seems impossible. However, the morphological difference between galagoes, lorises, and monkeys seems much greater than that between anthropoids and man. Furthermore, within the single family Cercopithecidae (old world monkeys) there seems to be greater morphological variation (e.g., between macaques, baboons, langurs) than there is morphological difference between man and the anthropoids.

As far as brain size and use of tools is concerned, there has been apparently a continuous line of development from the primitive hominids to modern man. Java man (? or ape-man) is so completely intermediate between South African man and modern man, and the difference between these two terminal forms so much a matter of degree, that it seems questionable whether these evolutionary stages justify generic separation. The difference is certainly not equivalent to a generic difference in most groups of animals.

17

Here we reprint another very influential paper by one of the founders and codifiers of the "modern synthesis" in evolutionary biology. G. G. Simpson was primarily a mammalian paleontologist, but also had strong interests in classification and taxonomy in general. In this paper he deals not with the content of hominid classifications or with human evolution *per se*, but with the way opinions or ideas about such matters can be expressed. His emphasis thus is on structure rather than content. One of Simpson's major points in this paper is that "much of the complexity and lack of agreement in nomenclature in this field does not, however, stem from ignorance or flouting of formal procedures but from differences of opinion that cannot be settled by rule or fiat." In other words, questions or controversies at the zoological/content level should not be confused with those at the purely nomenclatural/"linguistic" one. As pointed out in the Introduction, grouping specimens into taxa is inevitably subjective. Simpson makes the point that arguments at this level cannot be settled by application of the Code and that such disagreements are inevitable.

Simpson's suggestion (at the end of his section "Classification, Terminology, and Nomenclature") for improving specimen designation via using neutral site and specimen terminology rather than taxonomically suggestive genus and species names has been essentially universally adopted in paleoanthropology. Uniformity and consistency of reference was made much easier and more convenient by the comprehensive compilations of the *Catalogue of Fossil Hominids* (Oakley et al., 1971; 1975; 1977), published by the British Museum (Natural History).

Note Simpson's examples of how the same set of specimens can be assigned different sets of names, depending on a worker's underlying assumptions about species and evolution. Also note how the same name can have different implications when used in different contexts.

The Meaning of
Taxonomic Statements

George Gaylord Simpson

Introduction

Everyone who deals with evolution has occasion to use and to understand statements in the special language of taxonomy and classification. Communication is impeded by the facts that not all who use that language speak it fluently and that those fluent in it do not all speak the same dialect. In our conference on classification in relationship to human evolution we were talking this language much of the time. The main function of this contribution was to discuss the grammar and semantics of a reasonably standard dialect of the language. Centering the discussion on hominoid classification brings up and may clarify certain crucial points. This chapter is not, however, concerned with expressing opinions about human classification and evolution, but with discussing how such opinions are or should be expressed. I have recently covered theoretical aspects of animal taxonomy in some detail (Simpson, 1961), and mere repetition of parts of that book is here avoided.

Classification, Terminology, and Nomenclature

Taxonomic language involves not only a very large number of different designative words (names, terms) but also several different *kinds* of designations. The things or concepts designated by these words, technically their referents,[1] are also of different kinds, and the meanings or semantic implications are likewise diverse. It is therefore essential that they be clearly distinguished.

From George Gaylord Simpson, "The Meaning of Taxonomic Statements," in S. L. Washburn, Editor, *Classification and Human Evolution*, pp. 1–31, originally published in 1963 by Aldine and, subsequently, available from Books on Demand, University of Michigan.

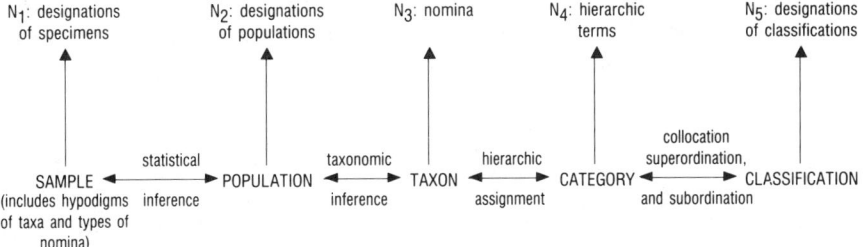

Figure 1.   Schema of processes (arrows), name sets (N), and referents (capitals) in taxonomy. Vertical arrows all represent the process of designation or symbolization. The processes represented by horizontal arrows proceed logically from left to right, but in practice no one operation can be carried out without reference to the others. These arrows are therefore drawn pointing both ways.

One way to do this is to consider the main operations involved in classification and the points or levels where special designations are required, as shown schematically in Figure 1.

The process starts with observation of the specimens in hand, the objective materials. The specimens studied and believed to be related in some biologically relevant way are a *sample*. If they are believed to represent a definite taxon (as determined at another level of inference), they constitute a *hypodigm*. Unequivocal designations of the specimens must refer to them as concrete, discrete objects; they are not designated by any name of the population or taxon to which they are supposed to belong. The ideal designation, practically universal in zoology but unfortunately not in anthropology, is by a collection or repository symbol and a catalogue number uniquely associated with each specimen. This is one kind of designation, one set of names (symbols of some sort, not necessarily or usually in words), and may be called the N_1 naming set.[2]

Observations and specimens, no matter how numerous, have no scientific significance purely per se. They acquire significance only when they are considered as representative of a larger group, or population, of possible observations or of individuals united by some common principle or relationship. The population may be abstract, for instance as symbolized in the equation for gravitation, applicable to a potentially infinite number of events but derived from a finite series of experimental observations. In zoological taxonomy the population is finite and concrete: a set of organisms existing (now or formerly) in nature. The existence and characteristics of that population are inferred from the sample drawn (we hope at random) from the population. The methods of inference are statistical by definition, which does not mean that any particular procedure of mathematical statistics is necessarily used although, of course, that is often appropriate and useful. A population is obviously not the same as the specimens actually studied, a sample drawn from the population.

At the next step in the process, all populations belong to taxa and all taxa are composed of populations. However, the two are not necessarily coextensive. It is often necessary to recognize and designate a local population that is a part of a taxon but does not in itself comprise a whole taxon. For some populations a different set of names or symbols, N_2, may therefore be required. Populations are in fact sometimes given distinct designations in zoological systematics, commonly by specification of their geographic location, but there is no established and uniform system. It may be sufficient to designate a population either as that from which a given sample was drawn (hence by extension of an N_1 designation) or as identical with that of a given taxon (hence by an N_3 designation).

A taxon is a group of real organisms recognized as a formal unit at any level of a hierarchic classification (Simpson, 1961, which see also for definitions and more extended discussions of hypodigms, categories, and hierarchies). A taxon is therefore a population, although the over-all population of one taxon may include many distinct populations of lesser scope. A taxon is created by inference that a population (itself statistically inferred from a sample which now becomes a hypodigm) meets a definition adopted for units in an author's classification. The set of designations for taxa, N_3 names, are those of formal, technical zoological nomenclature, e.g. *Homo*, the name of a taxon in primate classification. The word "name" is used in, many different ways, both in the vernacular and in technical discussion, and this has engendered confusion. I propose that technical Neo-Linnaean names in the N_3 set be called *nomina* (singular, *nomen*). Vernacular names ("lion," "monkey," "Neanderthal man") are in the N_3 set if they designate taxa, but they are not nomina.

Each taxon is assigned to (considered as a member of) a category, which has a defined rank in hierarchic classification. A category is a set, the members of which are all the taxa placed at a given level in such a classification. Categories are distinct from taxa, do not have populations as members, and are not represented by samples. They have their own set of names, N_4, which are the relatively few terms applied to levels of the Neo-Linnaean system: basically phylum, class, order, family, genus, and species, with various combinations in super-, sub-, and infra-, and occasionally such additional terms as cohort or tribe.

Finally the various taxa of assigned categorical rank are

collocated, superordinated, and subordinated among themselves and so form a hierarchic classification. This is done in terms of the N_3 (nomina) and N_4 (hierarchic terms) names. The added implications are conveyed less by nomenclatural than by topological means, primarily by arrangement and not consistently by verbal or related symbolization. Designations of classifications, N_5, are normally bibliographic references to their authors and places of publication.

What, now, are the meanings or implications of the various sets of designations? N_1 designations refer to particular objects. They imply only that a given specimen exists. They assure that when the same designation is used, the same object is meant. N_2 and N_3 designations both refer to groups that are considered to be populations related in some way. An author using such designations must make clear, explicitly or implicitly, the kind of relationship he has in mind. In modern zoology unless some other usage is definitely stated, it is generally understood that the relationship is genetic, that is, that it reflects evolutionary relationships. Concepts of what constitutes evolutionary relationships, how they are to be determined, and how reflected in classification become difficult and complex, but that is a different point.

Besides the implication that a population, usually genetic in relationship, is designated, nomina, N_3 names, further imply that the unit designated is given a definite rank in classification, that it is associated with an N_4 term. Under the International Code the forms of some nomina reflect the categorical rank of the corresponding taxa. For example, nomina ending in -idae (e.g. Hominidae) name families, and italicized, capitalized single words (e.g. *Homo*) name genera.

Most nomina, however, lack implications as to superordination, and *none have any implications beyond those mentioned*. For instance, nomina have no implications as to relationships among taxa at the same categorical level (e.g. *Homo* and *Tarsius*) or among taxa at any levels with etymologically distinct names (e.g. *Gorilla* and Pongidae). Further implications, which may be numerous and intricate as will be illustrated later, are inherent in the arrangement of nomina in a classification and not in the nomina themselves.

Discussion at the conference repeatedly illustrated the need for employing and distinguishing the different naming sets. The ambiguity and clumsiness of usual references to particular specimens and populations were especially evident. For example no clear and simple way was found for designating the various specimens from Olduvai Bed I that are believed not to belong to the taxon called *Zinjanthropus boisei* by Leakey. Presumably they will eventually be placed in taxa with distinct nomina (in the N_3 set), but that will not solve the problem of referring to the specimens themselves or to the populations inferred from them without ambiguity and without prejudice as to their taxonomic interpretation. As another example, no one maintains that *Telanthropus* is a valid taxon at the generic level, but no way has been found to refer to the specimens in question except as *Telanthropus*, an N_3 designation that necessarily implies a taxonomic conclusion agreed to be incorrect.

It must be emphasized that one of the greatest linguistic needs in this field is for clear, uniform, and distinct sets of N_1 and N_2 designations, applied to specimens and to local populations as distinct from taxa. Just what form such designations should take is a matter for proposal and agreement among those directly concerned with the specimens and their interpretation. It suffices here to stress that they *must not* have the form of Neo-Linnaean nomina. (The catalogue now being compiled by Oakley and Campbell may opportunely provide designations for specimens of fossil Hominidae.)

The Chaos of Anthropological Nomenclature

Men and all recent and fossil organisms pertinent to their affinities are animals, and the appropriate language for discussing their classification and relationships is that of animal taxonomy. When anthropologists have special purposes for which zoological taxonomic language is not appropriate, they should devise a separate language that does not duplicate any of the functions of this one and that does not permit confusion with its forms. There is, I believe, no reason for use of an additional language when what is being discussed is in fact the taxonomy of organisms. This language has been developed over a period of hundreds of years by cumulative experience and thought and has been thoroughly tested in nonanthropological use. It is admittedly imperfect, but for its purpose it is the best instrument available. Its imperfections call rather for improvement than for replacement. The most important needed improvement, with particular reference to anthropology, is that all those who use it should speak it well and in accordance with the best established current usages.

It is notorious that hominid nomenclature, particularly, has become chaotic. It is ironical that some of those who have most complained of the chaos have been leading contributors to it. A recent proposal that an international commission be formed to deal with the chaos refuses to recognize the appropriate code and the appropriate commission already set up. The author then proceeds to compound the confusion that he condemns.

Insofar as the chaos is merely formal or grammatical, it could be cleared up by knowledge of and adherence to the International Code of Zoological Nomenclature (Stoll et al., 1961), supplemented, if necessary, by whatever action might be proposed to and endorsed by the International Commission for Zoological Nomenclature. Much of the complexity and lack of agreement in nomenclature in this field does not, however, stem from ignorance or flouting of formal procedures but from differences of opinion that cannot be settled by rule or fiat. For example, when Leakey inferred from an Olduvai specimen (which he made a hypodigm) the existence of a taxon that he called *Zinjanthropus boisei* he was using correct taxonomic grammar to express the opinion that the taxon was distinct at both specific and generic categorical levels from any previously named. In equally grammatical expression of other opinions many other nomina, such as *Paranthropus boisei*, *Australopithecus robustus boisei*, or *Homo africanus boisei*, might have been proposed and might now be used. Or the specimen might have been and might now be referred to (or added to the hypodigm of) some previously named taxon such as *Paranthropus crassidens*. Any of those alternatives accord equally with the Code and would have equal status before the Commission. Decision among them is a zoological, not a nomenclatural or linguistic question, and it will be made by an eventual consensus of zoologists qualified in this special field.

Insofar as the chaos is due to faulty linguistics rather than to zoological disagreements, it stems either from ignorance or from refusal to follow rules and usages.[3] This must be almost the only field of science in which those who do not know and follow the established norms have so frequently had the temerity and opportunity to publish research that is, in this respect, incompetent.

An overt reason sometimes given for refusal to follow known nomenclatural norms is that some nomen is, in the opinion of a particular author, inappropriate. For example, some choose to rename *Australopithecus* as *Australanthropus*, thus adding another objective synonym to the chaos, on the grounds that the Greek *anthropos* more nearly expresses their opinion as to the affinities of the genus than does *pithekos*. The argument is completely irrelevant. *Australopithecus* does not mean "southern ape." Its meaning (defined by its referent) is simply the taxon to which the nomen was first attached and to which it was the first nomen attached. *Palaeolumbricus* or *Jitu* would have served just as well. The generic nomen does not, in itself, express any opinion as to the affinities of the taxon, and if nomina were changed in accord with every shade of opinion on affinities the chaos would be even worse than it is.[4]

Another reason for the chaos is the previously mentioned failure to develop and use consistently different designations for specimens, populations, and taxa, that is, distinct N_1, N_2, and N_3 name sets. A truly eminent anthropologist insisted on using the (N_3) nomen *Sinanthropus pekinensis* for specimens and a population although he concluded that this nomen does not designate a *taxon* specifically distinct from *Pithecanthropus erectus* or indeed from *Homo sapiens*. The example is far from unique.

Probably no one has ever admitted this, but it seems almost obvious that nomina (N_3) have sometimes been given to single specimens just to emphasize the importance of a discovery that could and should have been designated merely by a catalogue number (N_1). Of course no two specimens are alike, and it is always possible to fulfill the formal requirement that ostensible definition of a taxon must accompany proposal of a nomen. However, and again I would say obviously, the "definition" has often been only a description of an individual "type" with no regard for or even apparent consciousness of the fact that taxa are *populations*. This is not just a matter of exaggerating the taxonomic difference between specimens. It is a much more fundamental misunderstanding of what taxonomy is all about, of what nomina actually name. It is a relapse into pre-evolutionary typology, from which (I must confess) even the nonanthropological zoologists have not yet entirely freed themselves. Nomina have types, but not in the old typological sense. The types are not the referents of the N_3 nomina but are among the referents of N_1 designations. The referents of nomina are taxa—certain kinds of populations.

It is of course also true that the significance of differences between any two specimens has almost invariably come to be enormously exaggerated by one authority or another in this field. Here the fault is not so much lack of taxonomic grammar as lack of taxonomic common sense or experience. Many fossil hominids have been described and named by workers with no other experience in taxonomy. They have inevitably lacked the sense of balance and the interpretive skill of zoologists who have worked extensively on larger groups of animals. It must, however, be sadly noted that even broadly equipped zoologists often seem to lose their judgment if they work on hominids. Here factors of prestige, of personal involvement, of emotional investment rarely fail to affect the fully human scientist, although they hardly trouble the workers on, say, angleworms or dung beetles.

It is not really my intention to read an admonitory sermon to the anthropologists. You are all well aware of these shortcomings—in the work of others. I must pass on to matters more positive in value.

Species and Genera

The undue proliferation of specific and generic nomina is in part a semantic problem. The proposal of such nomina is rarely accompanied by an appropriate definition of the categories (as distinct from the taxa) involved, but ascribing specific or generic status to slightly variant specimens can be rationalized only on a typological basis. Whether consciously or not, taxa are evidently being defined as morphological types and statistical-taxonomic inferences from hypodigm to population to taxon (see Figure 1) are being omitted. But in modern biology taxa are populations and the following two nonconflicting definitions of the species are widely accepted:

Species are groups of actually or potentially interbreeding populations, which are reproductively isolated from other such groups.

An evolutionary species is a lineage (an ancestral-descendant sequence of populations) evolving separately from others and with its own unitary evolutionary role and tendencies. (Quoted from Simpson, 1961, where sources are cited and the definitions are further discussed.)

The naming of a species either should imply that the taxon is believed to correspond with one or both of those definitions or should be accompanied by the author's own equally clear alternative definition.

Evidence that the definition is met is largely morphological in most cases, especially for fossils. The most widely available and acceptable evidence is demonstration of a sufficient level of statistical confidence that a discontinuity exists *not* between specimens in hand but *between the populations inferred from those specimens*. The import of such evidence and the semantic implication of the word "species" are that populations placed in separate species are either

(1) in separate lineages (contemporaneous or not) between which significant interbreeding does not occur, or

(2) at successive stages in one lineage but with intervening evolutionary change of such magnitude that populations differ about as much as do contemporaneous species.

In dealing with the incomplete fossil record the information at hand commonly cannot establish the original presence or absence of a discontinuity. Allowance must be made for probabilities that further discovery will confirm or confute the existence of an ostensible discontinuity. Those probabilities depend on various circumstances. If populations are approximately contemporaneous, only moderately distinctive, and separated by a large geographic area from which no comparable specimens are known, there is considerable possibility that discovery of intervening populations would eliminate discontinuity. That is, for example, the situation regarding the original hypodigms of *Pithecanthropus erectus* and *Atlanthropus mauritanicus*. In my opinion the possibility that the Trinil population and the Ternifine population belong to the same species is such that different specific (a fortiori, generic) nomina are not justified at present.

If, on the other hand, populations being compared are of markedly different ages, decision to give them different specific nomina should depend on judgment whether such nomina would be justified if it turned out that they belong in successive segments of the same lineage. That would apply, for example, to the Mauer population as compared with the late Pleistocene European neanderthaloid population, and I should think would justify different specific nomina in this example.[5] Still a third situation arises when samples indicate populations that were approximately contemporaneous and living in the same region (synchronous and sympatric) as may be true, at least in part, for the Kromdraai, Swartkrans, Makapan, and Sterkfontein populations. In such cases allowance hardly has to be made for possible discoveries of populations living at other times and in different places. The degree of statistical confidence generated by the samples actually in hand may be taken as definitive of the probability of an original discontinuity, for instance between *Australopithecus africanus* and *A. robustus*.

The category genus is necessarily more arbitrary and less precise in definition than the species. A genus is a group of species believed to be more closely related among themselves than to any species placed in other genera. Pertinent morphological evidence is provided when a species differs less from another in the same genus than from any in another genus. When in fact only one species of a genus is known, that criterion is not available, and judgment may be based on differences comparable to those between accepted genera in the same general zoological group. There is no absolute criterion for the degree of difference to be called generic, and it is particularly here that experience and common sense are required.

It must be kept in mind that a genus is a *different* category from a species and that it is in principle a *group* of species. Much of the chaos in anthropological nomenclature has arisen from giving a different generic nomen to every supposed species, even some clearly not meriting specific rank. In effect no semantic distinction has been made between genus and species, and indeed the number of proposed generic nomina for hominids is much greater than the number of validly definable species. Monotypic genera are justified when, and only when, a single, isolated known species is so distinctive that the probability is that it belongs to a generic group of otherwise unknown ancestral, collateral, or descendent species. No one can reasonably doubt that this is true, for example, of *Oreopithecus bambolii* and that in this case the (at present) monotypic genus is justified. It is, however, hard to see how the application of more than one generic name to the various presently known australopithecine populations can possibly be justified, whatever the specific status of those populations may be.

Phylogeny and Resemblance

As most biologists understand modern taxonomic language, its implications are primarily evolutionary, but there is some persisting confusion even among professional taxonomists. It is not possible for classification directly to *express*, in all detail, opinions either as to phylogenetic relationships or as to degrees of resemblance. As a rule with important exceptions, degrees of resemblance tend to be correlated with degrees of evolutionary affinity. Resemblance provides important, but *not the only*, evidence of affinity. Classification can be made consistent with, even though not directly or fully expressive of, evolutionary affinity, and its language then has appropriate and understandable genetic implications. Classification cannot, at least in some cases, be made fully consistent with resemblance, and any implications as to resemblance are secondary and not necessarily reliable. These relationships can be explored by consideration of several hypothetical models or examples, set up so as to be simplified parallels of real problems in the use of taxonomic language to discuss human origins and relationships.

Classification and taxonomic discussion of related but distinct contemporaneous groups, such as the living apes and living men, involves a pattern of evolutionary divergence. That will first be discussed by means of a model. Discovery of related fossils almost always complicates the picture by revealing other groups divergent from both of those primarily concerned. It may, however, also reveal forms that are ancestral or that are close enough to the ancestry to strengthen inferences about the common ancestor and the course of evolution in the diverging lineages. In general the characters of two contemporaneous groups as compared with their common ancestry will tend to fall into the following classes, exemplified by characters of recent Pongidae and Hominidae:

A. Ancestral characters retained in both descendent groups. E.g. absence of external tail, pentadactylism, dental formula.

B. Ancestral characters retained in the first descendent group but divergently evolved in the second. E.g. quadrupedalism, grasping pes.

C. Ancestral characters retained in the second but divergent in the first group. E.g. undifferentiated lower premolars.

D. Characters divergently specialized in both. E.g. brachiation versus bipedalism.

E. Characters progressive but parallel in both. E.g. increase in average body size.

F. Convergent characters. I know of none between pongids and hominids, a fact which (if it is a fact) greatly simplifies judgment as to their affinities.

Different numbers of characters will fall into different categories. For instance in pongid-hominid comparison there are certainly many more A characters than any others and more B than C characters. (The given example of a C character is dubious.) Many characters do not simply and absolutely fall into one category or other. Retention of ancestral characters is usually relative and not absolute; some changes generally occur and "retained" usually means only "less changed." In constructing the simplest possible model on this basis, further simplifying postulates are that characters evolve at constant rates and that characters in the same group (e.g. D or E) evolve at the same rates. Those postulates are certainly never true in real phylogenies, and more realistic but also much more complicated models can be constructed by taking varying rates of evolution into account. The simplest possible limiting case, although unrealistic in detail, nevertheless more clearly illustrates valid and pertinent matters of principle. Such a model, analogous to pongid-hominid divergence, is illustrated in Figure 2. Numbers preceding the category designations symbolize relative numbers of characters in the corresponding categories. Exponents symbolize progressive change: a-b-c, or in a different direction x-y-z. It is assumed that in this example there are no F (convergent) characters. Roman numerals represent taxa: IV and V the two contemporaneous groups being compared, and I their common ancestry, ancestral to IV through II and to V through III.

From such data a comparison matrix can be formed. More sophisticated ways of doing this are exemplified in Campbell's contribution to this symposium, but for present purposes a simpler and sufficient method is to tabulate step differences between taxa. Change from C to C^a, for instance, is one step and from C^a to C^c

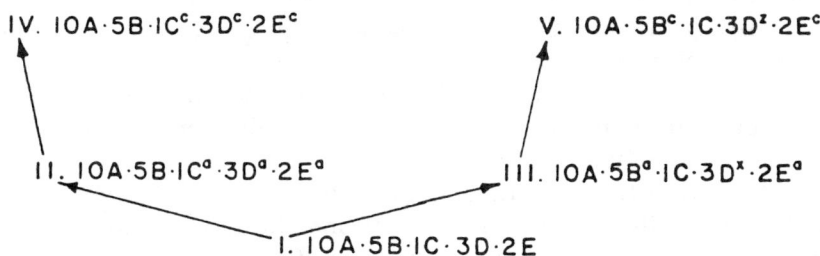

IV. $10A \cdot 5B \cdot 1C^c \cdot 3D^c \cdot 2E^c$

V. $10A \cdot 5B^c \cdot 1C \cdot 3D^z \cdot 2E^c$

II. $10A \cdot 5B \cdot 1C^q \cdot 3D^q \cdot 2E^q$

III. $10A \cdot 5B^q \cdot 1C \cdot 3D^x \cdot 2E^q$

I. $10A \cdot 5B \cdot 1C \cdot 3D \cdot 2E$

Figure 2. A model of simple evolutionary divergence. Symbols are explained in the text.

is two more. These are multiplied by the number of characters in the category, 1 for C characters. The matrix for the model in Figure 2 is given in Table 1. In this form of comparison, the smaller the number the greater the similarity. In this model I and II are most and IV and V least alike.

Let us suppose now that classification were to be based *entirely* on degrees of resemblance, as has been proposed by some taxonomists, and that classificatory language was therefore understood to be directly and solely expressive of resemblance. In building up higher taxa one would of course start by uniting I and II. If I and II are species, they would be placed in one genus; if genera, in one family. The maximum difference within the higher taxon would be 6. If no greater difference were allowed, all other lower taxa, III, IV, and V, would have to be placed in separate, monotypic higher taxa, an arrangement with nearly minimal significance, indicating no more than the close resemblance of I and II. If a difference of 12 were allowed in the higher taxon, II would be united with I and II, but IV should now also go with II, from which its difference is also 12. However, a taxon including IV and I would have to allow a difference of 18 and one including IV and III a difference of 24. But now V must also be added, for its difference from III is only 19.

Table 1

Comparison Matrix for Data of the Model in Fig. 2

	I	II	III	IV	V
I	0	6	9	18	29
II	6	0	12	12	32
III	9	12	0	24	19
IV	18	12	24	0	36
V	29	32	19	36	0

Thus *all* the lower taxa must go in a single taxon of next higher rank, an arrangement that indicates nothing of resemblances or relationships among any of those taxa. Insertion, or in actual examples discovery, of additional taxa, say between II and IV, would only compound the difficulties and lead still more inevitably to equally unsatisfactory alternatives.

I believe that the conclusion from the model is quite general for analogous real cases. In such situations the use of classificatory language as direct expression of degrees of resemblance commonly tends to produce one of two extremely inexpressive results: (1) one higher taxon includes the two most similar lower taxa and all other higher taxa are monotypic; or (2) one higher taxon includes all the lower taxa, no matter how numerous.[6]

Now let us agree that classificatory language is to have primarily evolutionary significance. For the moment degrees of resemblance need not be considered at all. It is clear from consideration of characters in categories B, C, and D that II can be ancestral to IV but not to III or V, and that III can be ancestral to V but not to II or IV. In actual instances the conclusions are neither so simple nor so obvious, but probabilities are readily established by the same categories of evidence. On this basis, II and IV can be placed in one higher taxon and III and V in another of the same categorical rank. That arrangement expresses the opinion, postulated as true in the model, that II and IV are phylogenetically related to each other and that III and V are also related in more or less the same way and degree. The arrangement is also consistent with but does not express the opinion, also postulated as true, that II is ancestral to IV and III to V.

In completion of this arrangement there are two alternatives as regards I. It could be placed in a third higher taxon ancestral to both of the two already formed, or it could be placed in the same higher taxon as II and IV, because it is phylogenetically closer to II than to III. Degree of resemblance here enters in as evidence for the latter inference.

Those are not the only classifications that would be consistent with the postulated evolutionary history. It would also be consistent to put I, II, and III in one higher taxon and IV and V in two others, or I, II, III, and IV in one and V in another. The implications on affinity would be somewhat different in each case but not conflicting: all are consistent with the postulates of the model. Choice would depend in part on what implications one wanted especially to bring out, since not all can be expressed in one classification. It would also depend on other considerations such as not changing previous classifications unnecessarily and conveying as much significant information as possible. (The last alternative mentioned above is the least informative.)

The model also illustrates the tendency, which is open to exception, for degree of resemblance to correlate with nearness of common ancestry. II and III are nearer their common ancestry than are IV and V, and they resemble each other more closely. The same is true of III and IV or of II and V as against IV and V. Such relationships are not directly implicit in the classification, but they are important in arriving at the judgments of affinity that are implicit in it.

Another important point illustrated in the model is that II and III resemble each other much more closely than III resembles its descendent V. It is realistic to expect an early—say Miocene— ancestor of *Homo* to be more like an ancestral ape than like modern man. It is unrealistic to expect the Miocene ancestors of either (or both) groups *necessarily* to have any of the specialized features that are diagnostic between *recent* members of the two families.

Taxonomic Language: Hominidae as Example

The Hominidae may be taken as an example of different principles (e.g. typological versus evolutionary) of classification and of the classificatory implications of different interpretations of data. For purposes of exemplification the data are postulated to be as in Figure 3A. Postulated ranges of variation of known specimens are indicated by the stippled areas, and in order to simplify the subsequent diagrams parts of those ranges are labeled A-F. X is a postulated individual specimen to be classified; it does not represent a specimen actually known. This arrangement is not presented as a realistic summary of what is, in fact, known. It is greatly simplified in several respects. Some known fossils do not fit clearly into the stippled areas, and some parts of those areas are not clearly represented by known fossils. Structure does not, in fact, follow a linear, one-dimensional scale and could be realistically indicated only by a (quite impractical) n-dimensional scale. Nevertheless this is the general *kind* of pattern, however grossly oversimplified, that the data do present.

Typological interpretation, Figure 3B, takes into account morphology only. It ignores temporal sequence and makes no phylogenetic interpretations. It abstracts an arbitrary number of fixed, distinctive types in the morphogenetic field and exercises subjective judgment as to whether a given, concrete specimen belongs to one type or another. Types may be hierarchically divided into subtypes, but variation is then ignored in the sub-types, and

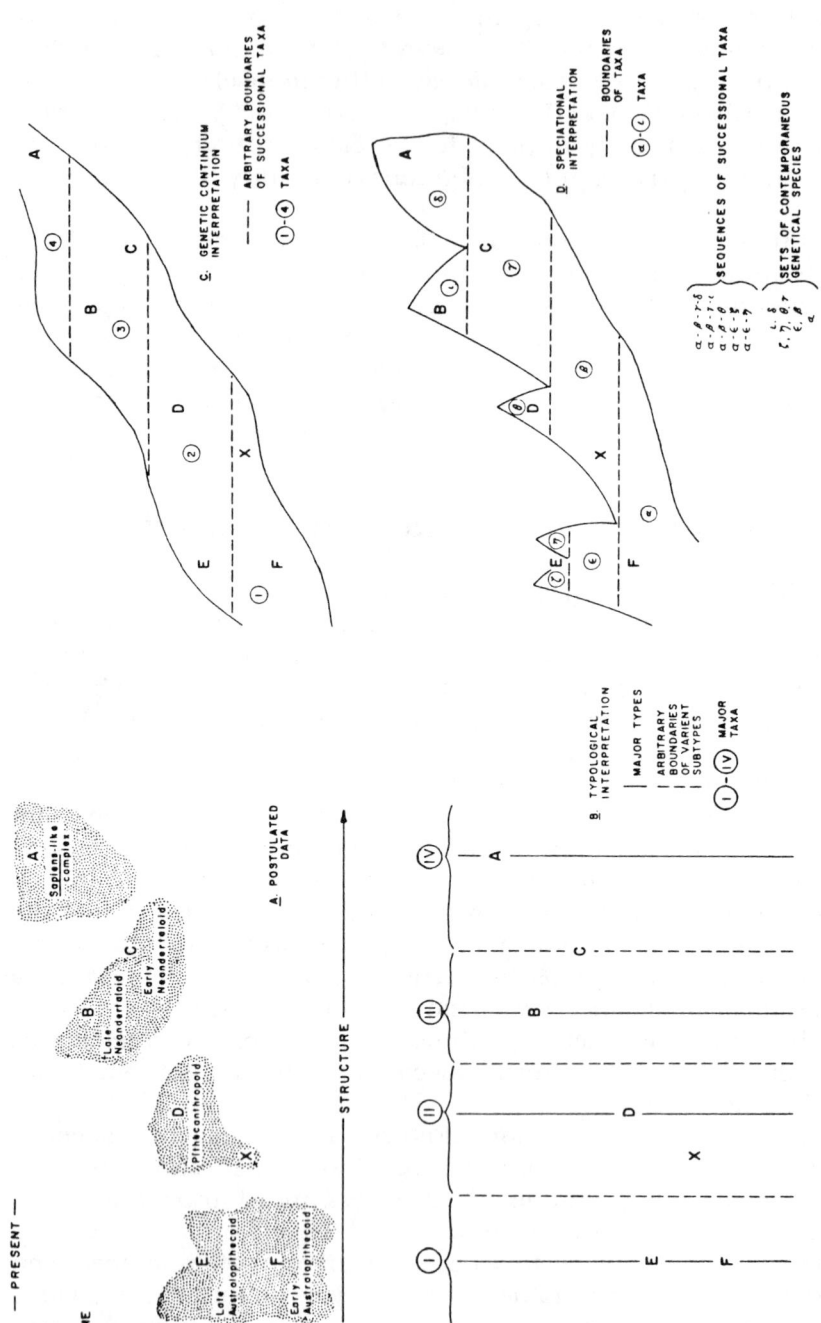

Figure 3.   Postulated data (simplified and generalized) and three possible taxonomic intrepretations of known hominids.   Further explanation in text.

genetical or evolutionary considerations do not enter in at any categorical level. As previously mentioned, this basis for classification has been largely abandoned in modern taxonomy. Nevertheless hominid classification started out on this basis, and even some of the most recent work in that field is at least covertly typological. The classificatory and nomenclatural expression of the typological arrangement in the diagram could take several forms, depending on the categorical level assigned to differences between types, for example:

I	*Homo africanus*	*Australopithecus africanus*
II	*Homo erectus*	*Pithecanthropus erectus*
III	*Homo neanderthalensis*	*Homo neanderthalensis*
IV	*Homo sapiens*	*Homo sapiens*

Figures 3*C* and *D* represent two possible kinds of phylogenetic interpretations of the same postulated data. Here temporal and genetic relationships are taken into account, and classification is based in principle on inferences as to evolutionary affinity. The differences between Figures 3*C* and *D* do not involve any difference in taxonomic principle but only in opinion as to probable evolutionary relationships. Both kinds of interpretation (with many differences in detail as to the particular placing of actual specimens) are currently supported by different students. Choice between them will depend on the accumulation of more data, and the ultimate arrangement will probably not be entirely of either kind in a clear-cut way but will involve elements of both. It will certainly be more complex than either of my diagrams.

Figure 3*C* diagrammatically represents the interpretation that hominids have been represented by only one interbreeding population since the early Pleistocene, at least. On this interpretation there is only one lineage or evolutionary species and only one genetical species at any one time. In that case, the species would have been highly variable, and even more so during much of past time than *Homo sapiens* is at present. At some time around the middle Pleistocene it might have varied all the way from what in purely morphological (or typological) terms could be called marginal australopithecoid through pithecanthropoid to marginal neanderthaloid. Such variation would be improbable within a single deme or local population. It would be less improbable among geographically separate (allopatric) populations or sub-species. Such geographic semi-isolates would of course be variable in themselves, but some might, for instance, vary about a more australopithecoid modal morphology and others about a more neanderthaloid mode. Discovery that fossil hominids fall into such modally distinct, synchronous but allopatric groups would favor this interpretation.

Whether current data do or do not tend to follow such a pattern I leave to the specialists in such matters.

The over-all ranges and modes of morphology change greatly from earlier to later parts of the phylogeny as postulated in Figure 3C, as they also do from early Pleistocene to now in the data actually known. It is useful, if not absolutely necessary, to take this into account in classification. The only possible way to do this (adhering to evolutionary taxonomic principles and accepting the interpretation as one genetic continuum) is to divide the lineage arbitrarily into successional taxa, as also exemplified in the diagram. The placing of the arbitrary boundaries and the ranks given the taxa will depend on judgment as to categorization of morphological differences and also, in practice, on where incomplete knowledge happens to make a morphological gap coincide more or less with a time line, as occurs between C and D in my postulated data. Again several different classifications would be consistent with the given phylogenetic interpretation, among them these two:

1. *Homo africanus*   *Australopithecus africanus*
2. *Homo erectus*   *Pithecanthropus erectus*
3. *Homo neanderthalensis*   *Homo neanderthalensis*
4. *Homo sapiens*   *Homo sapiens*

The same nomina are used here as in the typological interpretation, but the diagrams show that their meanings are different in the two. That is further shown by the fact that specimen X here falls into *africanus* but in Figure 3B into *erectus*.

Figure 3D represents an interpretation with speciation occurring within the Pleistocene hominid group so that there is not a single lineage but successive branching giving rise to two or more distinct, contemporaneous species, of which only one of the two last to arise has survived. The sets of contemporaneous species are separated by natural gaps (noninterbreeding) and are not arbitrary. Successive species in a lineage like α-β-γ-δ are arbitrary as regards that lineage, alone, but in the whole pattern their boundaries are also fixed by the (hypothetically) nonarbitrary points of splitting of the lineage into two species. The probability of this kind of pattern would be supported by discovery of contemporaneous (synchronous) populations with overlapping geographic distribution (sympatric) that did not intergrade and hence were probably not interbreeding significantly. (The existence of two or more distinct species does not, however, depend on their being sympatric.) Again I leave to the appropriate specialists whether data actually in hand do support such an interpretation.

One of several possible nomenclatures consistent with this pattern would be:

ζ *Australopithecus robustus*
η *Australopithecus africanus*
θ *Pithecanthropus erectus*
ι *Homo neanderthalensis*
δ *Homo sapiens*

It is not clear what actual specimens might fall into the hypothetical species α, β, γ, and ε, and I therefore suggest no nomenclature for them. It is clear, in any event, that some, at least, of the same nomina as used under the interpretations in Figure 3*B* and *C* are also applicable in *D* but again have different significance and contents. Specimen X, for example, is now neither in *africanus* nor in *erectus* but in unnamed hypothetical species β.

If identical nomina in Figures 3*C* and *D* referred to the same populations, there would be no ambiguity. Unfortunately, however, this is not likely to be the case. Population C in Figure 3*C* would probably be placed in the same taxon and referred to by the same specific nomen as population B. In Figure 3*D* population C would probably be placed in a different taxon and given a different nomen from population B. The ambiguity resides not in the taxonomic system but in the imperfection of our data and lack of agreement in their zoological interpretation. When such ambiguity persists, clarity demands that an author specify the populations included in his taxa, for example by adequate designation of their hypodigms. In the present example a possible clarifying device (if it accorded with an accepted zoological interpretation) would be to place populations B and C in separate subspecies. The placing of those subspecies in species would then clearly show different placing of the corresponding populations by different students.

Taxonomic Language: *Oreopithecus* as Example

The currently debated classification of *Oreopithecus* may be taken as another example of the use of classificatory language and its implications for phylogeny. Simply for purposes of the example, the following postulates are accepted:

1. Pongidae and Hominidae are distinct families of common ancestry and are united in the superfamily Hominoidea. This is now the usual conclusion and classification.

2. *Oreopithecus* had a common ancestry with both Pongidae and Hominidae at a time when the hominoid ancestry was distinct from that of any other recognized superfamily (e.g.

Cercopithecoidea). This is not established, but some, probably most, recent students consider it probable.

3. *Oreopithecus bambolii* is at least generically distinct from any other known species. This is universally accepted.

Table 2 gives four phylogenetic opinions (not the only ones possible) consistent with these postulates and gives for each two classifications consistent with that opinion and also consistent with further opinion as to the lesser or greater difference between *Oreopithecus* and forms considered allied to it according to the respective phylogenetic opinions.

Classifications B and C appear more than once in the table. Each is consistent with more than one opinion as to phylogeny. That is also true of all the other classifications, which are consistent with different phylogenetic opinions not distinguished in the table. For instance, A is consistent both with the view that *Oreopithecus* is directly ancestral to *Homo* and that it is a little-differentiated side branch from the direct ancestry. C appears three times in the table, but even that is not exhaustive: C is also consistent with several other possible opinions as to phylogeny. This forcefully illustrates

Table 2
Some Possible Opinions as to the Phylogeny and Distinctiveness
of *Oreopithecus*, and Some Classifications Consistent with Those Opinions

Opinions as to phylogeny of Oreopithecus	Opinions as to distinctiveness of Oreopithecus	
	a. Lesser	b. Greater
I. In or near hominine ancestry.	A. Hominoidea Hominidae Homininae *Oreopithecus*	B. Hominoidea Hominidae Oreopithecinae *Oreopithecus*
II. Divergent from early hominids after separation from pongids.	B. Hominoidea Hominidae Oreopithecinae *Oreopithecus*	C. Hominoidea Oreopithecidae *Oreopithecus*
III. Divergent from early pongids after separation from hominids.	D. Hominoidea Pongidae Oreopithecinae *Oreopithecus*	C. Hominoidea Oreopithecidae *Oreopithecus*
IV. Divergent from common stem of pongids and hominids.	C. Hominoidea Oreopithecidae *Oreopithecus*	E. Orepothecoidea Oreopithecidae *Oreopithecus*

the principle that classification is not intended to be an adequate expression of phylogeny but only to be consistent with conclusions as to evolutionary affinities.

Nevertheless each classification does have quite definite implications as to phylogeny; it is consistent with some and not with other opinions. Classification B, for instance, is definitely inconsistent with phylogenetic opinion III. In this sense a classification does express opinions on phylogeny in a broad or somewhat loose way. B does not express opinion as to whether *Oreopithecus* is a remote ancestor of *Homo*, an ancient hominid sidebranch, or a rapidly evolved hominine sidebranch, but it does express the opinion that *Oreopithecus* had a common ancestry with the hominids after hominids and pongids were distinct. C, which is the arrangement I personally prefer on present evidence, expresses the opinion that my postulate 2, above, is probable, and further that *Oreopithecus* is markedly differentiated from both pongids and hominids, but that the degree of distinctiveness is not so great as to warrant categorization as a superfamily. The basis for this preference is outlined in a following comment written after the conference, and most of the evidence is summarized in Straus's paper [appearing in same Washburn volume with this paper].

Conclusion

Classification is not an exact science and is not likely soon to become one. In order for its language to become completely unambiguous and uniform it would be necessary to have adequate samples of all pertinent populations, to reach universal agreement as to their affinities, and similarly to agree on just how to translate those affinities into formal classification. We do not have adequate samples and in the present field probably never will have. Even if the objective data were complete—indeed *especially* if they were complete—classification would have to have many arbitrary elements. Complete agreement as to what is to be said is thus chimerical and indeed may not even be desirable. We can, however, agree as to how to say what we mean. Taxonomic language as an instrument of communication has the failings but also the strengths of any other living language. Linguistic usages determine the clarity, not the content, of expression. The important thing is to use a language grammatically, to be quite sure of the implications of its words, and to make sure that it conveys our intended meanings unambiguously to others using the language.

Part Two: Added Notes

Aspects of Definition and Diagnosis

Attention has been given to the definition of man and his distinction, at various taxonomic levels, from his nonhuman or less fully human relatives. It may clarify matters to point out that the problem of definition is a complex with several different aspects. All are related and all intergrade, but ambiguity is introduced if no distinction is made among these aspects. The most essential distinctions may be illustrated by comparison of *Homo sapiens* and *Pan troglodytes*, the phylogenetic relationships of which are represented in Figure 4 as a simple splitting of one ancestral lineage into two and the subsequent divergent evolution of the latter into the present terminal species named. That there were also other

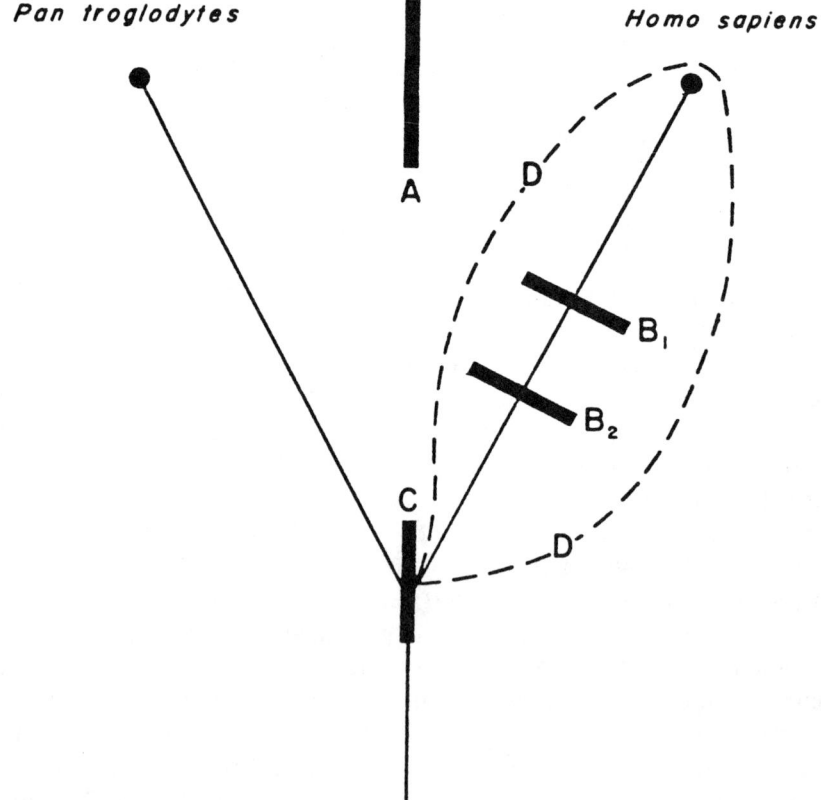

Figure 4.   Diagram of different aspects of diagnosis between chimpanzee and man, and between living and fossil man. Further explanation in text.

splittings and other lineages is a complication that can be ignored for purposes of the example.

The terminal species are completely discrete and, as living populations, completely accessible biological and taxonomic units. The diagnosis of living man with respect to the chimpanzee, symbolized by A in the diagram, is simple and unambiguous, and it may be made in almost any terms we like, for example by the presence of different proteins in the blood, by the nonoverlapping frequency distributions of premolar structure or brain size, by the differences in locomotion, or by the absence in one and presence in the other of true language. Diagnosis of *Homo sapiens* with respect to his own ancestry, represented by B in the diagram, will theoretically involve some but not all of the same criteria as the A-diagnosis. It will involve such diagnostic characters as evolved within the human ancestry since the separation from the chimpanzee lineage, but will exclude characters of the common ancestry preserved in *Homo* and not in *Pan*. An operational B-diagnosis must also exclude characters not determinable in the available fossils, because it is the fossils, only, to which the diagnosis must be applied in practice. The origin of such A-diagnosis characters as language, self-awareness, human social structure and the like is of greatest evolutionary interest, but simply is not pertinent to the practical classification of actual fossil specimens.

In principle a B-diagnosis involves drawing a more or less arbitrary line across what was in nature a continuum. In practice these lines are commonly placed where there are gaps in knowledge. There are at present no known specimens unequivocally recognizable as intermediate between *Homo sapiens* and *H. erectus* or between *Homo* and *Australopithecus*, but it is reasonably certain that such creatures existed and probable that specimens of them will eventually be found. In the meantime our ignorance of them permits unambiguous diagnoses between those specific and generic taxa. Campbell (this book) has pointed out that successive taxa may be naturally separated at peaks of evolutionary rates, and it is also likely that such peaks may correspond with gaps in present knowledge. At present, however, and for purposes of practical classification we cannot definitely place such peaks or even be quite sure that they did exist, and in any event a *precise* diagnostic boundary would still be an arbitrary line in a continuum.

It is clear that different characters and character-complexes (e.g. the locomotory, masticatory, and central nervous system complexes) have evolved at different rates and accelerated at different times. The possible B-diagnoses are therefore multiple and will involve different characters or complexes at different times, as symbolized by the separation of B_1 and B_2 in the diagram. As total change is gradual and cumulative, it will also be ranked at different categorical levels at different points. Such ranking is not completely arbitrary because it can be more or less coordinated with current categorical ranking of taxa separated by A-diagnosis. That agreement on such ranking can be reached is evident by unanimity in the contributions to this book that *Homo sapiens* and *H. erectus* differ only specifically but that *Australopithecus africanus* is generically distinct from both. Some disagreement exists at higher levels, but only by one hierarchic substep, e.g. as between super-family and family or family and subfamily.

Another aspect of definition is symbolized by C in the diagram, representing the difference between the ancestors of *Pan* and *Homo* when they first separated. At the very time of the separation, this distinction was (with high probability if not complete certainty) at the specific level only. As Dobzhansky pointed out, it is further likely that the initially differentiating characters of *both* groups occurred together as simple variants in a single species ancestral to the two. The actual characters involved are likely to be (in simplest, incipient form) some but certainly not all of those eventually involved in the A-diagnosis. I can see no reliable way of judging a priori just which of those characters do in fact stem from the C-differentiation, and indeed it is possible in principle that the C-diagnosis will be by characters not present in either *Pan* or *Homo*. The C-diagnosis does not depend on the characters entirely in themselves or on the magnitude of the difference at the time of separation, but on our knowledge (or opinion) as to the taxonomic significance of the *subsequent* divergence of the lineages involved. The categorical level is assigned *ex post facto*: in the Miocene our ancestor was probably one of several species of a genus of Pongidae but now that same species, if definitely identified, might properly be classified in a quite distinct genus of Hominidae.

Finally, it is desirable that the whole evolutionary unit enclosed in the broken line marked D in the diagram should also be recognized and named as a taxon. Its definition would ideally involve all the differentiating aspects of A, B, and C.

The Dentition of Oreopithecus

The dentition of *Oreopithecus*, which is completely known from a large number of specimens and which I have been able to study in detail through the courtesy of Dr. J. Hürzeler, is extremely distinctive. It differs more from either pongid or hominid dentitions than living pongids and hominids, together, differ among themselves. The proportions of canines and premolars are indeed hominid-like, but the morphology of those teeth, which I take to be far more significant than a simple matter of proportions, is decidedly nonhominid, as is also that of all the other teeth. I completely agree with Dr. Hürzeler's conclusion, strongly reinforced by the study presented in this book by Dr. William L. Straus, Jr., that *Oreopithecus* is not a cercopithecoid and that it is morphologically a hominoid. I cannot, however, support reference to the Hominidae. This negative conclusion is based not only on the great differences of the teeth of *Oreopithecus* from those of any sure hominid (the genera *Homo* and *Australopithecus*, both *sensu lato*, or known Homininae and Australopithecinae) but also on the following considerations:

1. There are known forms (notably *Ramapithecus* and *Kenyapithecus*) approximately contemporaneous with *Oreopithecus* that have much more hominid-like dentitions.

2. The peculiarities of the *Oreopithecus* cheek teeth (e.g., specialized protoconule region and course of the crista obliqua on the upper molars; presence of a mesoconid and of a peculiar paraconid or, more likely, pseudoparaconid on the lower molars) are all present in incipient or fully developed form in *Apidium*, an early Oligocene genus. (Dr. E. L. Simons, and earlier Dr. W. K. Gregory have pointed out this resemblance, which is also mentioned in Straus's paper in this book. Dr. Simons has several very important new specimens of *Apidium* which have not yet been described in print but which I have seen through his courtesy.)

3. No Anthropoidea have yet been definitely identified before the early Oligocene. Comparison of all known Anthropoidea and of conceivably ancestral earlier prosiminians strongly suggests that the anthropoid ancestry in general and the hominoid ancestry in particular lacked peculiarities of dentition already present in *Apidium* and emphasized in *Oreopithecus*.

These considerations seem to me to indicate that the lineage leading to *Oreopithecus* was already distinct near the very base of hominoid differentiation and that its dental differences from both pongids and hominids are not on the whole primitive but are at least in large part divergent with respect to the ancestry of those groups. If that is correct, the placing of *Oreopithecus* (and *Apidium*) in a family Oreopithecidae of the superfamily Hominoidea, for which a preference is expressed above, becomes almost obligatory. It is further highly improbable that the important skeletal resemblances of *Oreopithecus* to the Hominidae, as listed by Straus, are primitive for the Hominoidea. They would therefore appear to be the result of parallelism or convergence to a degree that is noteworthy but not inherently improbable. The total result is a peculiar adaptive type quite unlike that of either the Pongidae or the Hominidae.

This example also illustrates two points of general taxonomic principle. First, the varying significance of different characters and character complexes for classification requires that they be weighted in the light of the whole biological and evolutionary picture as far as it is known. Second, there are often, as in this case, considerations highly pertinent to biological classification and yet difficult or impossible to reduce to simple numerical form and to include in a computer program for obtaining a (likewise highly pertinent) coefficient of similarity or distance function.

Affinities and Classification of the Hominoidea

Within the last few years data for classification of the hominoids have been greatly enriched in breadth and depth: discovery of new pertinent fossils; continued anatomical investigation; studies of serology, hemoglobins, and chromosomes; more detailed behavioral observations of nonhuman hominoids, particularly under natural conditions. Since almost every conceivable view (along with some rather inconceivable ones) has been upheld at one time or another, this new information is useful not so much in giving us a new pattern of affinities as in enabling us to choose more surely among the many already proposed and to gain more confidence in various points of detail. Without reviewing evidence or arguments, in this note I shall briefly state how the probabilities look to me now. I shall also briefly sketch some of the main, different family-group (superfamily, family, subfamily) classifications that would be consistent with those probabilities and indicate my own preference among them.

Gibbons.—On fossil evidence, the gibbon ancestry was probably distinct from that of other apes when hominoids first appear in the fossil record, early Oligocene, and was certainly so in the Miocene. Recent karyological and serological evidence, presented in this book by Klinger and by Goodman, respectively, also indicates strong divergence of living gibbons from all other living hominoids. No evidence suggests special affinities with orangs, on one hand, or with the chimpanzee-gorilla group on the other. Special affinity with *Homo* is out of the question. It is probable that these three groups did not diverge among themselves until after the gibbon ancestry had already split off. On the other hand, the gibbons have not diverged radically from other apes either morphologically or adaptively. What is distinctive in their facies is largely due to their having remained smaller than other hominoids and to their specialized locomotion, which in turn seems to require the first peculiarity. Miocene fossils (demonstrated to members of the 1962 conference by Professor Zapfe in Vienna) suggest that the locomotory specialization evolved comparatively late. Although gibbons, strictly speaking, and siamangs are usually placed in different genera, they are manifestly very closely related and I now prefer to place them all in *Hylobates*.

Orangs.—Morphologically and to some extent also adaptively *Pongo* is not markedly unlike the living chimpanzees and, to less extent, gorillas. This has long, although not quite unanimously, been considered evidence of rather close relationship. Schultz, whose knowledge of orangs is unexcelled, continued to uphold that view in the conference. On the other hand, karyological (Klinger) and serological (Goodman) evidence seems to separate *Pongo* from *Pan* (and *Gorilla*) as sharply as *Hylobates*. Fossil orangs have not been identified before the Pleistocene, but there is no evident reason why the ancestry of *Pongo* may not be found near that of *Pan* in the dryopithecine complex. On balance, it still seems probable that *Pongo* is especially related to the African apes, but that the split was far enough back to permit considerable, more or less clandestine molecular and chromosomal divergence. Morphological divergence has been less, probably because of retention of somewhat similar adaptation.

Gigantopithecus.—When known only from isolated molars, this Chinese Pleistocene genus was claimed to be a hominid. Later finds of lower jaws and dentitions, not yet adequately described as far as I know, seem clearly to exclude it from the Hominidae. It seems to be a terminal specialization not very close to any living form. During the 1962 conference Leakey suggested special affinity with *Pongo*, and that is a possibility. On present very inadequate evidence I would, however, prefer to place it only as Pongidae *incertae sedis*, and I omit it from further consideration.

African apes.—A consensus has always considered gorillas and chimpanzees as especially and rather closely related, and all the recent evidence, including that of serology and karyology, confirms that view. They are of course sharply distinct species, at a point of divergence where experienced taxonomists may well waver between giving only specific or also generic weight to that divergence. Merely listing characters that demonstrate the self-evident fact of their distinctness does not necessarily suffice to maintain the time-honored generic separation, and at present I prefer to consider both chimpanzees and gorillas as species of *Pan*. Whether *P. paniscus* is a valid third species, closer to *P. troglodytes* than to *P. gorilla*, is still moot. Placing all the African apes in *Pan* permits classification to express the clear fact that they are much more closely related to each other than to any species of other genera, and henceforth I shall use the nomen *Pan* in this sense.

It has long been the virtually universal opinion that *Pan* is anatomically and adaptively rather close first to *Pongo* and then to *Hylobates*. Recent studies, while also confirming that these are quite distinct groups well separated at a generic level, at least, agree with the old conclusion that the three genera belong together in a natural taxon at some higher level. Nevertheless, as noted above, newer subanatomical evidence suggests that separation of the ancestors of the three genera within that higher taxon is ancient. The situation is complicated only by comparisons with the Hominidae, summarized below.

No explicit and particular connection of *Pan* with a Tertiary ancestry has yet been found or, at least, clearly recognized. It is, however, probable that in a general sense the ancestry occurred somewhere in known or unknown members of what is here called the dryopithecine complex.

The dryopithecine complex.—The Miocene and Pliocene of Africa, Europe, and Asia have produced many specimens clearly apelike and distinct from contemporaneous closer relatives of *Hylobates* (notably *Pliopithecus* and the closely allied, perhaps not generically distinct, *Limnopithecus*). They are otherwise highly diverse and clearly represent a greater number of lineages than the four or perhaps five recent species that might possibly have arisen from this complex (*Pongo pygmaeus*, *Pan troglodytes*, perhaps *Pan paniscus*, *Pan gorilla*, and *Homo sapiens*). Many or most of the dryopithecine-complex lineages have therefore become extinct, and it is the opposite of surprising to find that some of them (e.g. *Proconsul*) have combinations of characters not found in taxa as diagnosed primarily on the basis of living species.

With the sole exception of *Proconsul*, the members of this complex are known only from very incomplete remains, largely single teeth or unassociated upper and lower jaws with partial dentitions. Their classification and nomenclature are unsatisfactory and almost chaotic within the group. This situation could surely be improved by a revision even of the already known fragments, and I consider such revision plus a really systematic search for better specimens the greatest desideratum of primate paleontology at present. Some of these forms, such as *Dryopithecus* itself, may be rather near the ancestry of the living great apes, *Pan* and perhaps also *Pongo*. Others, such as *Ramapithecus*, and possibly *Kenyapithecus*, may belong near the ancestry of *Homo*. Still others, as already mentioned, are doubtless more or less terminal in lineages not close to any living forms. If or when more probable affinities with later groups are established, it should be possible to place some of the dryopithecine-complex species and genera in taxa, e.g. in subfamilies, currently based primarily on living species. It is, however, my opinion that the present unsatisfactory stage of study and incompleteness of sampling do not establish such connections at a sufficient level of probability.

Oreopithecus.—Elsewhere in this chapter I have sufficiently expressed the opinion that *Oreopithecus* probably represents a lineage separate from near the base of hominoid differentiation, with limited parallelism with the Hominidae, but culminating in an extinct terminal form adaptively very unlike either the great apes or any hominids.

Australopithecus.—Despite earlier polemics, it is now perfectly clear that among other sufficiently known genera *Australopithecus* (*sensu lato*, including *Paranthropus* and *Zinjanthropus*) is most closely allied to *Homo*. Late *Australopithecus*, at least, is almost certainly contemporaneous with early *Homo* (including *Pithecanthropus*, etc.) and hence not ancestral to it. Present evidence does not exclude, and may be taken to favor, the possibility that early *Australopithecus*, or an unknown genus close to it, was such a direct ancestor. Although *Australopithecus* greatly strengthens the opinion that *Homo* had an apelike ancestor in common with the living great apes, it does not at present seem to me to give additional clear evidence as to precisely which apes, among living forms or the dryopithecine complex, are most nearly related to *Homo*.

Homo.—Since the 19th century it has been the usual, although by no means the universal, opinion that among living mammals *Homo* is most closely allied to *Pan*. That conclusion, based originally on classical anatomical grounds, is strongly supported by all the new evidence, anatomical, karyological, biochemical, and behavioral, presented at recent conferences. It now seems to me so probable that other alternatives need no longer be seriously considered. It should, however, be strongly emphasized that *Homo* represents an anatomical and adaptive complex very radically different from that of any other known animal and (with the partial exception of *Homo*'s close ally *Australopithecus*) differing far more from other living or adequately known fossil hominoids than they differ among themselves. Seemingly contradictory evidence (e.g. that of the haemoglobins as reported by Zuckerkandl [appearing in same Washburn volume with this paper]) indicates merely that in *certain* characters *Homo* and its allies retain ancestral resemblances and that *these* are not the characters involved in their otherwise radical divergence—a common and indeed universal phenomenon of evolution.

Affinities, adaptive radiation, and phylogeny.—Figure 5 shows, by combination of a dendrogram and an adaptive grid, my present views as to the affinities and the adaptive or structural-functional relationships of the living hominoids. Interpretation of probable closeness of genetic connection is indicated by depth of branching, although it is to be emphasized that such a diagram is not a phylogenetic tree and has no time dimension. Adaptive or ecological (and corresponding structural-functional-behavioral) resemblances and differences are approximated by horizontal distances between the terminal points.

Figure 6 shows in schematic form the combined phylogenetic inferences reviewed above for various groups of hominoids.

Classification.—Evolutionary classification takes into account: degrees of homologous resemblance in *all* available respects; the most probable phylogenetic inferences from all data (including the foregoing resemblances plus evolutionary analysis and weighting of the various characteristics); and also the practical needs of discussion and communication.

ADAPTIVE AND STRUCTURAL-FUNCTIONAL ZONES

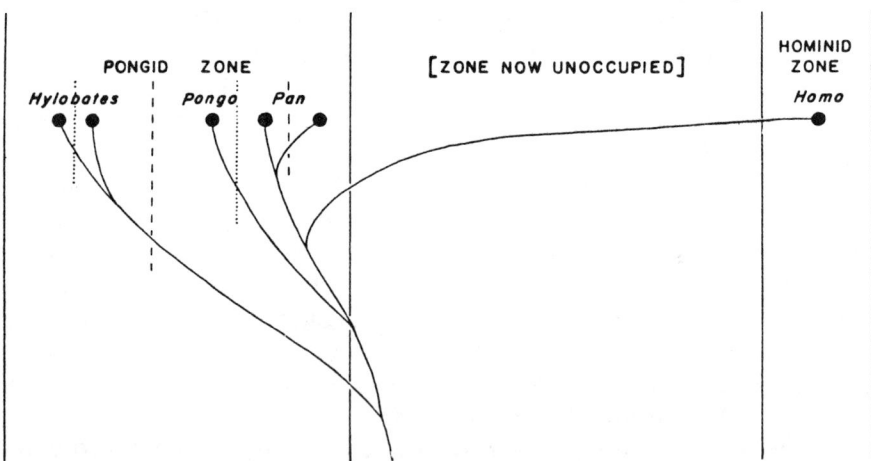

Figure 5.   Dendrogram of probable affinities of recent hominoids in relationship to their radiation into adaptive-structural-functional zones. The two major adaptive zones are bordered by solid lines. Pongid radiation into sub- and sub-sub-zones is schematically suggested by broken and dotted lines. A dendrogram of this sort has no time dimension and does not indicate lineages, but it is probable that divergences of lines showing affinities are topologically similar to the phylogenetic lineage pattern.

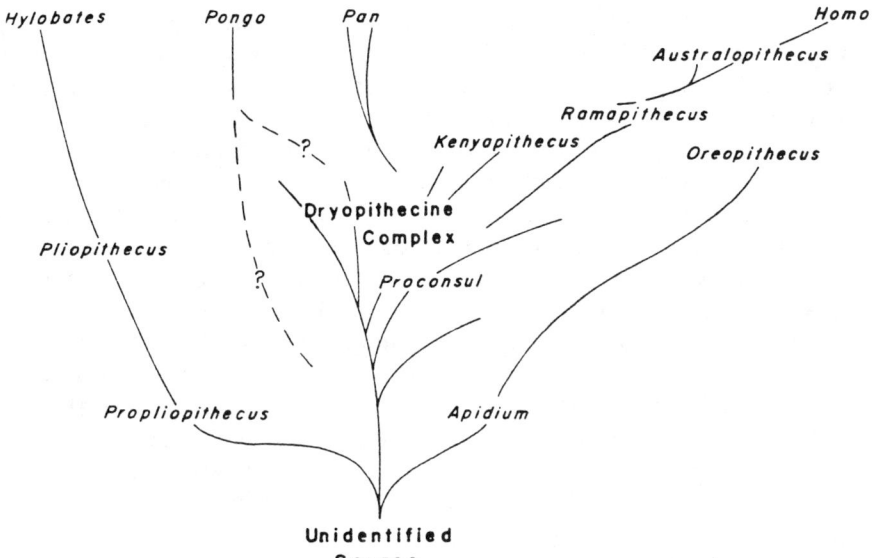

Figure 6.   Tentative and schematic phylogeny of the Hominoidea. Most of the individual fossil genera are omitted, and lineages as drawn are meant to be impressionistic and diagrammatic (especially in and around the dryopithecine complex) rather than representing all or particular generic or specific lines.

It now seems perfectly clear and is all but universally recognized that the animals here called hominoids (in anticipation of a conclusion) form a natural evolutionary unit that should be recognized and named as a taxon. When the whole order Primates is taken into account, the categorical level of this taxon should clearly be no higher than infraorder and no lower than family. A case could be made out for either extreme of those rankings, but in my opinion the intermediate ranking, that of superfamily, is best in balance and convenience. It also accords with the recent consensus, and thus with the principle that communication is best served if nomenclature is not changed unnecessarily. The current and nomenclaturally correct nomen for this superfamily is Hominoidea.

At the next lower level, I have already expressed the opinion that the apparent lineage *Apidium-Oreopithecus* is at present best ranked as a family, because of its ancient separation plus its marked divergence from any other group now usually given family rank. In view especially of Straus's analysis [appearing in same Washburn volume with this paper], the only reasonable alternative is to rank this lineage as a subfamily of Hominidae on balance of anatomical resemblance alone. That is not wholly excluded by present evidence, but its phylogenetic implications seem to me extremely improbable. (Unless *Pongo*, *Pan*, and all the dryopithecine complex were also placed in the Hominidae it would be definitely inconsistent with the phylogeny of Figure 6.)

Because of the clear and ancient separation of *Hylobates* and its fossil allies from other hominoids, those forms are now frequently, probably usually, given family status. On the other hand, *Hylobates* almost certainly had a common hominoid ancestry with *Pongo* and *Pan*, and its evolutionary divergence from those genera and their fossil allies is decidedly less than that of either *Homo* or *Oreopithecus*. That would justify placing the *Hylobates* group as a subfamily of a family also containing *Pongo*, *Pan*, and some, at least, of the dryopithecine complex. Both arrangements are consistent with reasonable interpretations of the available data, and choice becomes a matter of personal judgment and convenience. I continue to prefer the second alternative, partly as a matter of linguistic convenience. One frequently wants to distinguish humans and apes (plus or including gibbons) and this is most conveniently done at the family level. The secondary distinction between gibbons and (other) apes is convenient at the subfamily level.

If the gibbons are given family rank, an analogous argument can be made for also giving *Pongo* separate family rank, since it cannot be *demonstrated* to have split off more recently from the *Pan* ancestry and since it now proves to be serologically about equally distinct. This is nevertheless still largely an argument from ignorance, and the most extensive positive evidence we have, that of anatomy, still suggests closer affinities between *Pongo* and *Pan*. Certainly if *Hylobates* and *Pan* are in one family, *Pongo* belongs in the same family and the alternatives involve subfamilies. Of the five possible subfamily arrangements of the three genera (and their fossil allies), only placing *Hylobates* in one subfamily and *Pongo* and *Pan* in another or placing each in a separate subfamily seem worthy of consideration on present knowledge. Both can be defended, but I continue tentatively to favor the former, because I believe that *Pongo* and *Pan* probably are more closely related than *Hylobates* and *Pongo* and because I think monotypic subfamilies should be as few as possible.

For the dryopithecine complex, there are three possibilities: (1) all could be put in the same subfamily as *Pongo* and *Pan*; (2) they could all be placed in a subfamily or family of their own; (3) those clearly related to *Pongo*, *Pan*, or *Homo* could be put in family-group taxa with those genera and the others in one or more separate subfamilies or families. The first could be justified on grounds of general resemblance and of probability that *part* of this complex is near the ancestry of *Pan*, perhaps also of *Pongo*. The second arrangement makes a horizontal grouping that is phyletically complex and to some extent artificial, but is justifiable *faute de mieux* in our present lack of almost any good knowledge of detailed relationships in this group, and pending withdrawal of particular genera if their affinities with other established taxa are later demonstrated. The third arrangement is definitely preferable and should ultimately be adopted, but as indicated above present knowledge seems inadequate to follow it with sufficient probability. I now waver between (1) and (2), but hesitantly continue to follow (2) simply because there seem to be insufficient grounds for changing it until this complex is better understood.

It is now virtually established that the affinities of *Homo* with *Pan* are closer than with *Pongo* or *Hylobates*. That suggests the possible desirability of placing *Pan* in the Hominidae and *Pongo* and *Hylobates* in one or two other families, an arrangement supported by Goodman and by Klinger at the conference. In fact, as noted above, this view as to affinities is an old one and has been held for two or three generations by students who nevertheless all excluded *Pan* from the Hominidae. The new data increase confidence in the conclusion as to affinities but do not, in my opinion, either require or justify the proposed change in classification. The question involves the whole complex of taxonomic principles and cannot be argued in detail here. The following, in briefest form, are among the principal reasons for continuing to exclude *Pan* from the Hominidae:

1. *Pan* is the terminus of a conservative lineage, retaining in a general way an anatomical and adaptive facies common to all recent hominoids *except Homo* (and probably to all adequately known fossil ones except *Australopithecus* and the very different *Oreopithecus*). *Homo* is both anatomically and adaptively the most radically distinctive of all hominoids, divergent to a degree considered familial by all primatologists.

2. *Pan* is obviously not ancestral to *Homo*. The common ancestor was almost certainly more *Pan*-like than *Homo*-like, which suggests not that *Pan* should go in the Hominidae but that the common ancestor should be in a separate family with *Pan* (or in still another ancestral family).

3. When a younger family arises from an older, the situation is frequently or usually similar to that of the *Pan* and *Homo* lineages: one of several lineages of the older family splits into two or more, *one* of which diverges (and/or diversifies) until its descendents warrant family status. If the lineages that did *not* diverge are also placed in the later family on the basis of more recent common ancestry, carrying this process

on down will eventually require inclusion of all descendents of earlier splittings also in the latest family—eventually the whole animal kingdom would be in the Hominidae on this principle. An arbitrary division must be made in practical classification, and the obvious place to make it is where the lineage *later* reaching family status split off— in this instance where the ancestors of *Homo* split from those of *Pan* and other pongids. Classification cannot be based on recency of common ancestry *alone*.

4. Both arrangements are equally consistent with our present understanding of hominid phylogeny, but the proposed new arrangement is less consistent with other evolutionary considerations, notably that of adaptive divergence. Therefore the change is neither required nor warranted. That radical change of nomenclatural usage would also create great confusion in discussion of hominoid relationships.

Subfamily separation of *Australopithecus* and *Homo* became usual at the time when placing of *Australopithecus* in the Pongidae (as here used) or Hominidae (also of present usage) was disputed and the australopithecines were claimed to include several genera. Now that there is essential agreement that *Australopithecus* belongs in the Hominidae, I see no sufficient reason for having two subfamilies, especially as each has only one known genus as I and, I believe, most others now define the genera. "Australopithecine" and "hominine" may still be used as strictly vernacular terms for structural levels, although there is little need for such terms as long as we know only one genus at each level.

Finally, Leakey suggested at the conference a possible alternative classification that requires comment. (I do not know whether he himself considers the alternative preferable and proposes to use it.) He pointed out that the Hominoidea of my classification could be reduced to family rank and that my families and subfamilies, with some reassignment of genera, could be considered subfamilies (plus another subfamily for *Proconsul*, which I give only generic rank). The supporting phylogeny looked rather different from mine, but was topologically almost identical. Aside from minor points, either classification was consistent with either phylogeny. Without going into detail, I here only say that at the very best I do not think his classification an improvement justifying so many departures from current usage. The reasons are mostly implicit or explicit in the preceding comments.

The following is the outline classification that I now favor, with all accepted recent genera included but a number of fossil genera omitted:

Superfamily Hominoidea
 Family Pongidae
  Subfamily Hylobatinae

Pliopithecus
Hylobates
Subfamily Dryopithecinae
Dryopithecus and other genera of the dryopithecine complex, *tentatively* including *Proconsul*, *Ramapithecus*, and *Kenyapithecus*.
Subfamily Ponginae
Pongo
Pan
Family Hominidae
Australopithecus
Homo
Family Oreopithecidae
Apidium
Oreopithecus

(Placing Oreopithecidae at the end and next to Hominidae has no special significance; as Darwin noted over a century ago, you cannot put organisms in one linear evolutionary sequence or show their true affinities on a sheet of paper. Putting *Australopithecus* in the Hominidae and adding the Oreopithecidae are the only essential changes from my 1945 classification, which I believe *in general* to be consistent with the mass of more recent information. Unfortunately that is less true of some of the nonhominoid primates.)

Notes

[1] A psycholinguistic term also useful in zoological taxonomy. See Brown (1958).

[2] Recognizing different sets of taxonomic designations and distinguishing them in this way is due to Gregg (1954), although I do not follow him in detail.

[3] Mayr, Linsley, and Usinger (1953) provide an excellent introduction to the rules and basic taxonomic usages. The promulgation of a later code (Stoll *et al.*, 1961) must, however, now be taken in account.

[4] It is true that when the system was being developed, from 200 to 250 years ago, the then relatively few nomina were usually intended to be etymologically descriptive. The experience of two centuries has, however, conclusively demonstrated that as a general principle this is absolutely unworkable. Except for the occasional mnemonic value, it is unfortunate that nomina do often have ostensible etymological meanings in addition to their real, taxonomic meanings.

[5] I am not suggesting what those nomina should be. Among many other possibilities they might be *Homo heidelbergensis* and *Homo neanderthalensis*, or *Homo erectus* and *Homo sapiens*.

[6] Even extremists who would classify by resemblance *only*, usually admit that the biological significance of such classification may be confused by differential rates of evolution and by convergence. Note that in our simplified model both of those admitted sources of confusion have been eliminated by postulation, and that biologically significant classification from the numerical data alone still is impossible.

Bibliography

Brown, R. 1958. *Words and things.* Glencoe, IL: Free Press.

Clark, W. E. Le G. 1959. *The antecedents of man.* Chicago: Quadrangle Books.

Cold Spring Harbor Symposia on Quantitative Zoology. 1950. "Origin and evolution of man." *C. S. H. Symposia Vol. XV.* [Especially Mayr, pp. 109–118.]

Genovés, T. S. 1960. "Primate taxonomy and *Oreopithecus.*" *Science, 133*:760–761. [A recent example of misunderstanding of taxonomic language, corrected by Straus (1960).]

Gregg, J. R. 1954. *The language of taxonomy.* New York: Columbia Univ. Press.

Heberer, G. 1961. "Abstammung des Menschen." In, Bertalanffy and Gessner, *Handbuch der Biologie.* [Incomplete; classification, disagreeing with my concepts of taxonomic language, especially in Lieferung 117/118, pp. 287, 307.]

Mayr, E., Ed. 1957. *The species problem.* Amer. Assoc. Adv. Sci., Pub. No. 50.

Mayr, E., E. G. Linsley, and R. L. Usinger. 1953. *Methods and principles of systematic zoology.* New York: McGraw-Hill.

Simons, E. L. 1959. An anthropoid frontal bone from the Fayum Oligocene of Egypt: The oldest skull fragment of a higher primate. Amer. Mus. Novitates, No. 1976.

Simpson, G. G. 1945. "The principles of classification and a classification of mammals." Bull. *Amer. Mus. Nat. Hist., 85*:i–xvi, 1–350. [Primate classification now requiring much updating but still illustrative of the method.]

_____. 1961. *Principles of animal taxonomy.* New York: Columbia Univ. Press.

_____. 1962. *Primate taxonomy and recent studies of nonhuman primates.* New York Acad. Sci., Conference on Relatives of Man. [Contains material pertinent to the present topic and not repeated here.]

Stoll, N. R., et al. 1961. *International code of zoological nomenclature.* London: Internat. Trust for Zool. Nomencl.

Straus, W. L., Jr. 1960. "Primate taxonomy and *Oreopithecus.*" *Science, 133*:760–761. [Correction of misuse of taxonomic language by Genovés (1960).]

18

Bernard Campbell's comprehensive review of hominid nomenclature up to the time of its original publication (1965) is an indispensable source and reference on this subject. He reviews the literature of hominid taxonomy, compiling synonymies of names and bibliographic references. He attempts to establish which names are available and which are invalid for a variety of reasons. This paper can be seen as continuing the critiques of Mayr and Simpson (selections 16 and 17) of previous taxonomic practice in paleoanthropology by providing a convenient source of type specimens and the names assigned to them.

Campbell expresses his goal as follows: ". . . we have separate named species for almost every fossil find of any importance from the Pleistocene, a state of affairs which does not in any way reflect the actual taxonomy of evolving man. It is not so much that early workers were 'splitters,' in the taxonomic sense, but that they were ignorant of the meaning of the concept of species, and used binomial nomenclature as a system of labeling. This publication is prepared as a guide in the current movement which aims at a revision of the classification of the Hominidae so that its structure reflects as nearly as possible the reality of evolving populations as they replaced each other during the Pleistocene."

The Nomenclature of the Hominidae

Bernard G. Campbell

1. Introduction

The need for a review of the nomenclature of the fossil Hominidae—the taxonomic family which includes man and his immediate family—has been apparent for some time and has been urged by many authorities. The family presents problems in its nomenclature for a number of reasons: proximity to their subject has led many workers to attach too elevated a significance to individual fossil discoveries; the resulting proliferation of names and arbitrary multiplication of species and genera has produced confusion in classification. Animal taxonomy lies within the province of the zoologist; many Hominid fossils have been described and named by archaeologists or anatomists who are either not fully acquainted with or who have ignored standard taxonomic practice.[1] Innate problems arising from the inadequate methodology of taxonomy are particularly apparent in a group so closely studied as this.

A full review of the nomenclature of the family, here attempted for the first time, has been made imperative by the rapid increase in the amount of fossil material which has recently occurred. The aim of the present work is to resolve the existing confusion in Hominid nomenclature. It is the intention to examine objectively the latin names, or nomina[2] proposed for fossil Hominid finds in the light of the International Code of Zoological Nomenclature, in

From Bernard G. Campbell, *The Nomenclature of the Hominidae*, Royal Anthropological Institute, Occasional Paper no. 22, pp. 1–33 (1965). Reprinted by permission of the Royal Anthropological Institute of Great Britain and Ireland. The three indices originally included with this paper have been omitted in this reprinted version.

order to establish which of the numerous existing *nomina* are invalid and may be removed from the literature, and which are available for the designation of Hominid taxa. Any reviewer who wishes to construct a revised classification of the family will find in the present work the correct *nomen* to use for any taxon which warrants a specific or subspecific rank. If the reviser feels that no separate taxon (be it genus, species, or subspecies) is justified on the basis of a particular named fossil, there is no need to retain the *nomen*, which becomes a *subjective junior synonym*.

This distinction between the two necessary stages in arriving at a classification cannot be overstressed. The correct *nomen* for a taxon is an objective fact—not a matter of taste, as many suppose. The arrangement of the taxa in a classification, on the other hand, is an entirely subjective assessment of relationship. The contents of this paper can be claimed to be noncontroversial, therefore, and its conclusions reliable, subject, of course, to human errors and omissions.

The object, therefore, in bringing together the data for the nomenclature of the Hominidae is to narrow the field of taxonomic controversy by obviating semantic misunderstanding. Disagreement may be inevitable, but it is to be hoped that misunderstanding is not. Communication between human palaeontologists is necessary for advance in knowledge. This review of nomenclature is an attempt to improve that communication.

The method adopted will be to list all the Pleistocene fossils which have been placed in the Hominidae and which have received *nomina*, with those names that they have been given, signifying whether each is valid, and if not why, and whether it has priority. They are not listed in the form of a formal classification but simply under the site names by which the fossils are well known. It is of course impossible to ensure that the list of proposed names is comprehensive, since they have been published in many countries and in diverse journals. Nevertheless it is felt that all those names have been included which have any part to play in the nomenclature of the group.

Post-glacial fossils have not been included in this work, although very many have been made the types of new taxa. All such taxa are considered to be objective junior synonyms of *Homo sapiens*. The compilation of these taxa would be onerous, and since they are not essential for preparing a classification of the Hominidae, they have been omitted from this work. It has been ascertained, however, that the species-group *nomina* which have priority for the major subspecies of living man are as follows:

H. sapiens sapiens L. 1758.

H. sapiens afer L. 1758.

H. sapiens asiaticus L. 1758.

H. sapiens americanus L. 1758.[3]

H. sapiens australasicus Bory de St. Vincent, 1825.

H. sapiens neptunianus Bory de St. Vincent, 1825.[4]

The Code of Zoological Nomenclature

Although the rules were drawn up expressly to promote stability in zoological nomenclature, in practice they have been found frequently to demand changes which instil anything but stability into the nomenclature of the group under study. This is because a large percentage of animal taxa were named before the Code was first drawn up on an international scale (1889), though even in those early days, the injunctions of Linnaeus were often ignored. Since 1889, an equally large percentage of animal taxa have been described without regard to the rules and recommendations of the various International Zoological Congresses. The result of this situation is that the original rules have been increasingly modified to the extent of making them extremely complex. A revised edition of the International Code of Zoological Nomenclature has now been published (1961) and has been used as the basis of this survey.[5] An explanation of the interpretation which has been placed upon certain articles of the Code in evaluating Hominid taxonomy is included in the notes to the second section of this paper.

The Type Concept

The concept of the *type* in taxonomy is vital to a consistent application of the binomial system of nomenclature. Today the concept of the type carries no special biological significance, but is necessary for an unambiguous designation of a taxon. In particular a difficulty in nomenclature arises where a mixed population of fossils is given a name without a type being specified, and is later classified as two groups. Under these conditions there is no *a priori* means of deciding which of the resulting taxa retains the original name.

New taxa are very rarely presented in palaeoanthropology according to the rules of the more recent zoological congresses. Where only a single individual is involved, no problem exists, and this individual is considered to constitute the *holotype*, and is here indicated as such.[6] Where more than one individual is involved, and none is specified as holotype by the author, they must all be considered *syntypes*. When this occurs, the reviser of the group is at liberty to select a type (the lectotype)[7] in order to avoid further confusion, and this course has been followed here. The choice of the lectotype has been based on the condition of the fossil, and on the typicality of its morphology, amongst the group of syntypes from which it is taken. Those syntypes which are not selected then lose their status as types in the taxon.[8]

The selection of a lectotype in this way avoids any subjective judgement at this stage on the homogeneity or otherwise of the populations involved. In most instances, however, the first individual from a site to be described with a new binomen is taken to be the holotype of the taxon.

The Law of Priority

This well-known law in taxonomy states that 'The valid name of a taxon is the oldest available name applied to it provided that the name is not invalidated by any provision of this Code' (CINZ, Article 23). *Nomina* published subsequently, are not considered to be invalid but are retained as *objective junior synonyms*, or more generally, *synonyms*. In order to include all valid latin names and to put them on record, a synonymy is included in this catalogue for all fossils which have been named more than once. It is hoped that such a record will help to avoid confusion in the future.

Some Reasons for the Non-availability of Nomina

Where *nomina* have been published which have proved to be invalid, this fact is indicated in the following lists by letters. It is necessary at this stage to list those reasons which are most commonly met with which render names invalid.

(*a*) *Generic Names.* A generic name may be invalid because it is a *homonym* or a *nomen nudum*. A generic homonym is not available for the erection of a new genus since it has already been used for this purpose. It is said to be occupied (CINZ, Articles 52 and 53). In certain cases it is not clear whether the author intended to place a species in a pre-existing genus of hominids, or supposed that he was creating a new one. (Palaeoanthropologists rarely follow the recommendations of the International Zoological Congresses—printed in Appendix E of CINZ—which require that a new name be stated to be new.) Generic names are treated as homonyms in this paper therefore where the author's intention would appear to be to create a new name, even if it is not indicated as such. Homonyms can rather easily arise where generic names are created from a limited number of Greek and Latin roots meaning ancient, first, early, half or ape—roots usually attached to the Greek word *anthropos*, meaning *man*. Many such names were suggested by Haeckel for hypothetical groups and were later used for naming real specimens. A generic homonym in this paper is indicated by the letters **GH**—in heavy type.

A generic name may also be a *nomen nudum*. This occurs when the specimen to which it has been given is not sufficiently well identified. A generic name introduced before 1931 may be invalid if it is not accompanied by an 'indication, definition, or description' (CINZ, Articles 12 and 16). Such a name is indicated by letters **GA**. Furthermore a generic name introduced after 1930 may be invalid if it is not also accompanied by a statement in which the author 'purports to give characters differentiating the taxon' (CINZ, Article 13(*a*)) or is not accompanied by the citation of a type-species (Article 13(*b*)). Such names are indicated by letters **GB**.

(*b*) *Specific Names*. Specific homonyms are specific *or subspecific* names which have been previously used for the designation of a species or subspecies within the same genus (CINZ, Articles 52 and 53). Such names are indicated by the letters **SH**.

A specific name may also be a *nomen nudum*. Articles 12 and 16, together with Article 13(*a*) apply here as they do to generic names. *Nomina nuda* arising under Articles 12 and 16 before 1931 are indicated by the letters **SM**. Those arising under Article 13(*a*) are indicated by the letters **SN**.

There is a third possibility here, however, that a new name 'proposed conditionally, or one proposed explicitly as the name of a "variety" or "form" is not available' if it was proposed after 1960 (Article 15). Such a name is indicated by the letters **SO**.

Presentation of Type-Specimens of the Hominidae

Specimens which have been made the types of named taxa in the family Hominidae are not listed under their latin names, but under the names of the sites in which the fossils have been found. Since these site names are so well known, it is believed that this arrangement will be more convenient.

Under each site heading, the following information is given:

1. SITE NAME, Country, and date of discovery.
2. *Nomina* given to fossils with author, date and full reference in each case. That name which has priority in every case precedes the remainder. These latter are either objective junior synonyms or invalid names.

3. Type fossil and reference to a description of it. The reference is not given in full where this is the same paper as that in which was published the latin name.
4. Reference is included here to any further description of the type which may prove useful.
   The sites are listed under continents.

2. List of Named Hominid Taxa

Africa

1. BOSKOP, S. Africa 1913.
2. **Homo capensis**, Broom 1917.
   Homo sapiens Boskop, Gregory 1921. Invalid **SM**.[9]
   R. Broom 1917, Fossil Man in South Africa, *Amer. Mus. J.,* **17**, 141–2.
   W. K. Gregory 1921, *The Origin and Evolution of the Human Dentition*, Baltimore.
3. Holotype: calvaria of adult female (?) in S. H. Haughton 1917, Preliminary Note on the Ancient Human Skull-Remains from the Transvaal, *Trans. Roy. Soc. South Africa*, **6**, 1–14.

1. BROKEN HILL, Zambia 1921.
2. **Homo rhodesiensis**, Woodward 1921.
   Cyphanthropus rhodesiensis, (Woodward 1921) Pycraft 1928.[10]
   Homo primigenius africanus, Weidenreich 1928. **SH**.
   A. S. Woodward 1921, A new cave man from Rhodesia, South Africa, *Nature Lond.*, **108**, 371–2.
   W. P. Pycraft 1928, Description of Human Remains, in *Rhodesian Man and Associated Remains*, British Museum (Natural History), London, pp. 1–51.
   F. Weidenreich 1928, Entwicklungs- und Rassentypen des *Homo primigenius*, *Natur u. Mus.*, **58**, 1–13 and 51–62.
3. Holotype: skull of adult male, in A. S. Woodward 1921.

1. CAPE FLATS, S. Africa 1929.
2. **Homo drennani**, Kleinschmidt 1931.
   Homo australoideus africanus, Drennan 1929. **SH**.
   O. Kleinschmidt 1931, *Der Urmensch*, Leipsig.
   M. R. Drennan 1929, An Australoid Skull from the Cape Flats, *J. Roy. Anthrop. Inst.*, **59**, 147–427.
3. Lectotype: skull described in M. Drennan 1929.

1. EYASI, Tanzania 1935.
2. **Palaeoanthropus njarasensis**, Reck & Kohl-Larsen 1936.[11]
   Africanthropus njarasensis, (Reck & Kohl-Larsen 1936) Weinert 1938. **GH**.
   H. Reck & L. Kohl-Larsen 1936, Erster Uberblick über die jungdiluvialen Tier- und Menschenfunde Dr Kohl-Larsen's im nordoslichen Teil des Njarasa-Grabens (Ostafrika), *Geol. Runsch.*, **27**, 401–41.
   H. Weinert 1938, Der erste afrikanische Affenmensch, *Africanthropus njarasensis*, *Der Biologie*, **7**, 125.
3. Holotype: skull, as reconstructed, in H. Reck & L. Kohl-Larsen 1936.

1. FLORISBAD, S. Africa 1932.
2. **Homo (Africanthropus) helmei**, Dreyer 1935.
   Homo florisbadensis, Drennan 1935 (post-dates Dreyer 1935).
   T. F. Dreyer 1935, A Human Skull from Florisbad, Orange Free State, with a Note on the Endocranial Cast (by C. U. Ariëns-Kappers), *Proc. Acad. Sci. Amsterdam*, **38**, 119–28.
   M. R. Drennan 1935, The Florisbad Skull, *S. African J. Sci.*, **32**, 601–2.
3. Holotype: calvaria, in T. F. Dreyer 1935.

1. GARUSI, Tanganyika 1939.
2. **Meganthropus africanus**, Weinert 1950.
   Präanthropus, Hennig 1948. Invalid **GB**.
   H. Weinert 1950, Uber die neuen Vor- und Fruhmenschenfunde aus Afrika, Java, China und Frankreich, *Z. Morph. Anthrop.*, **42**, 113–48.
   E. Hennig 1948, Quartärfaunen und Urgeschichte Ostafrikas, *Naturwiss. Rdsch.*, **1**, 5.
3. Holotype: Maxillary fragment, in H. Weinert 1950.
4. A. Remane 1951, Die Zahne des *Meganthropus africanus*, *Z. Morph. Anthrop.*, **43**, 311–29.

1. HOPEFIELD (Saldanha), S. Africa 1953.
2. **Homo saldanesis**, Drennan 1955.
   M. R. Drennan 1955, The special features and status of the Saldanha Skull, *Am. J. Phys. Anthrop.*, **13**, 625–34.
3. Holotype: Reconstructed calotte and mandible fragment, in M. Drennan 1955.
4. R. Singer 1954, The Saldanha Skull from Hopefield, South Africa, *Am. J. Phys. Anthrop.*, **12**, 345–62.

1. KANAM, Kenya 1932.
2. **Homo kanamensis**, Leakey 1935.
   L. S. B. Leakey 1935, *The Stone Age Races of Kenya*, London, 9–24.
3. Holotype: mandibular fragment, in L. S. B. Leakey 1935.

1. KANJERA, Kenya 1932.
2. **Homo leakeyi**, Paterson 1940. Invalid **SN**.
   T. T. Paterson 1940, Geology and Early Man, *Nature Lond.*, **146**, 12–15, 49–52.
3. _____
4. L. S. B. Leakey 1935, *The Stone Age Races of Kenya*, London.

1. KROMDRAAI, S. Africa 1938–41.
2. **Paranthropus robustus**, Broom 1938.
   R. Broom 1938, Pleistocene Anthropoid Apes of South Africa, *Nature Lond.*, **142**, 377–9.
3. Holotype: skull and mandible fragments, in R. Broom & G. Schepers 1946, The South African Fossil Ape Men—The *Australopithecinae*, *Trans. Mus. Mem.*, **2**, 84–113.

1. MAKAPANSGAT, S. Africa 1947–49.
2. **Australopithecus prometheus**, Dart 1948.
   R. A. Dart 1948, The Makapansgat Protohuman *Australopithecus prometheus*, *Amer. J. Phys. Anthrop.*, **6**, 259–83.
3. Holotype: occipital bone, in R. A. Dart 1948.

1. OLDUVAI, Tanzania 1959.
2. **Zinjanthropus boisei**, Leakey 1959.
   L. S. B. Leakey 1959, A new fossil skull from Olduvai, *Nature Lond.*, **184**, 491–3.
3. Holotype: Cranium, in L. S. B. Leakey 1959.
4. P. V. Tobias 1966 (in press).

1. OLDUVAI, Tanzania 1960(*a*).
2. **Homo leakeyi**, Heberer 1963.
   G. Heberer 1963, Uber einen neuen archanthropinen Typus aus der Oldoway-Schlucht, *Z. Morph. Anthrop.*, **53**, 171–7.
3. Holotype: Calotte of adult male from Olduvai Bed II, in L. S. B. Leakey 1961. New finds at Olduvai Gorge, *Nature Lond.*, **189**, 649–50.
4. P. V. Tobias, *in preparation*.

1. OLDUVAI, Tanzania 1960(*b*).
2. **Homo habilis**, Leakey, Tobias & Napier 1964.
   L. S. B. Leakey, P. V. Tobias & J. R. Napier 1964, A new species of the genus *Homo* from Olduvai Gorge, *Nature Lond.*, **202**, 7–9.
3. Holotype: Mandible with dentition, upper molar, parietals and handbones of juvenile from site FLKNN. I, Olduvai Bed I, in L. S. B. Leakey 1961, New Finds at Olduvai Gorge, *Nature*, **189**, 649–50, *and* The Juvenile Mandible from Olduvai, *Nature*, **191**, 417–18.
4. J. R. Napier 1962, Fossil Hand Bones from Olduvai Gorge, *Nature*, **196**, 409–11.
   P. V. Tobias 1964, The Olduvai Bed I Hominine with special reference to its cranial capacity, *Nature*, **202**, 3–4.
   P. V. Tobias, *in preparation*.

1. STERKFONTEIN, S. Africa 1936.
2. **Australopithecus transvaalensis**, Broom 1936.
   Plesianthropus transvaalensis, Broom 1938.
   R. Broom 1936, A New Fossil Anthropoid Skull from Sterkfontein, near Krugersdorp, South Africa, *Nature Lond.*, **138**, 486–8.
   R. Broom 1938, Pleistocene Anthropoid Apes of South Africa, *Nature Lond.*, **142**, 377–9.
3. Holotype: skull base, maxilla, teeth and endocranial cast, in R. Broom 1936.[12]

1. SWARTKRANS, S. Africa 1949.
2. **Paranthropus crassidens**, Broom 1949.
   R. Broom 1949, Another New Type of Fossil Ape-Man (*Paranthropus crassidens*), *Nature Lond.*, **163**, 57.
3. Holotype: mandibular fragment with five teeth, in R. Broom 1949.
4. R. Broom & J. T. Robinson 1952, Swartkrans Ape-Man: *Paranthropus crassidens*, *Transvaal Mus. Mem.*, **6**, 1–123.

1. SWARTKRANS, S. Africa 1949.
2. **Telanthropus capensis**, Broom & Robinson 1949.
   R. Broom & J. T. Robinson 1949, New Type of Fossil Man, *Nature Lond.*, **164**, 322–3.
3. Holotype: 'Telanthropus I' mandible, in R. Broom & J. T. Robinson 1949.
4. R. Broom & J. T. Robinson 1950, Man Contemporaneous with the Swartkrans Ape-Man, *Am. J. Phys. Anthrop.*, **8**, 151–5.

R. Broom & J. T. Robinson 1952, Swartkrans Ape-Man: *Paranthropus crassidens*, *Trans. Mus. Mem.*, **6**, 1–123.

1. TAUNG, S. Africa 1924.
2. **Australopithecus africanus**, Dart 1925.
   R. A. Dart 1925, *Australopithecus africanus*: The Man-Ape of South Africa, *Nature Lond.*, **115**, 195–9.
3. Holotype: skull of child of about 6 years, in R. A. Dart 1925.
4. R. Broom & G. W. H. Schepers 1946, The South African Fossil Ape-Men—The *Australopithecinae*, *Trans. Mus. Mem.*, **2**, 1–272.

1. TERNIFINE, Algeria 1954.
2. **Atlanthroprus mauritanicus**, Arambourg 1954.
   Homo (pithecanthropus) atlanticus, Dolinar-Osole 1956.
   Homo (pithecanthropus) ternifinus, Dolinar-Osole 1956.
   C. Arambourg 1954, L'hominien fossile de Ternifine (Algerie), *C. R. Acad. Sci. Paris*, **239**, 893–5.
   Z. Dolinar-Osole 1956, Nova Pitekantropoidna oblika hominida iz severne Afrike, *Arheoloski Vestnik*, **7**, 173–80.
3. Holotype: mandible of adult 'Atlanthropus I,' in C. Arambourg 1955, A recent discovery in human paleontology: Atlanthropus of Ternifine (Algeria), *Amer. J. Phys. Anthrop.*, **13**, 191–6.
4. C. Arambourg 1963, Le gisement de Ternifine, *Arch. Inst. Pal. Hum. Mem.*, **32**.

1. YAYO, Chad 1961.
2. *Tchadanthropus uxoris*, Coppens 1965. Invalid **SO**.
   Y. Coppens, L'Hominien du Tchad, *C. R. Acad. Sci. Paris*, **260**, 2869–71.
3._____

Asia

1. CHENCHIAWO (Lantian County, Shensi Province), China 1963.
2. **Sinanthropus lantianensis**, Woo 1964.
   J. K. Woo 1964, Mandible of *Sinanthropus lantianensis*, *Current Anthrop.* **5**, 98–101.
3. Holotype: mandible, in J. K. Woo 1964.

1. CHOUKOUTIEN, China 1927.
2. **Sinanthropus pekinensis**, Black & Zdansky 1927. (In D. Black 1927.)
   Pithecanthropus sinensis, Weinert 1931.
   Praehomo asiaticus sinensis, Eikstedt 1932.
   Sinanthropus (Pithecanthropus?) pekingensis, Hennig 1932.[13] **SH**.
   D. Black 1927, On a Lower Molar Hominid Tooth from Chou-Kou-Tien Deposit, *Palaeontologia Sinica*, Ser. D, **7**, 1–28.
   H. Weinert 1931, Der 'Sinanthropus pekinensis' als Bestätigung des Pithecanthropus erectus, *Z. Morph. Anthrop.*, **29**, 159–87.
   E. Eickstedt 1932, Hominiden und Simioiden Über den derzeitigen Stand der Abstammungsfrage, *Z. artzl. Fortbild.*, **29**, 608–13.
   E. Hennig 1932, Fortschritte der Altsteinzeit-Forschung in der Alten Welt, *Petermann's Mitt.*, **78**, 134–7.
3. Holotype: Lower left molar, in D. Black 1927.

1. DJEBEL KAFZEH, Israel 1934–5.
2. (*Homo semiprimigenius palestinus* Montandon, 1943.)[14]
   G. Montandon 1943, *L'homme Préhistorique et les Préhumains*, Paris.
3. _____
4. H. V. Vallois, *in preparation*.
   M. Boule & H. V. Vallois 1952, *Les Hommes Fossiles. Elements de Paléontologie Humaine*, 4th ed., Paris, 394–5.

1. HONG KONG, 1935.
2. **Sinanthropus officinalis**, von Koenigswald 1952.
   G. H. R. von Koenigswald 1952, *Gigantopithecus blacki* von Koenigswald, a Giant Fossil Hominoid from the Pleistocene of Southern China, *Anthrop. Papers Amer. Mus. Nat. Hist.*, **43**, Part 4, 291–326.
3. Holotype: Right Upper M1, in G. von Koenigswald 1952.

1. HONG KONG, 1935–9.
2. **Hemianthropus peii**, von Koenigswald 1957, **GH**,[15] Invalid **SN**. (Ref. 1957*a*.)
   Hemanthropus peii, von Koenigswald 1957. Invalid **GB**, **SN**. (Ref. 1957*b*.)
   G. H. R. von Koenigswald 1957*a*, Remarks on Gigantopithecus and other Hominoid remains from

Southern China, *Proc. Kon. Ned. Acad. Wetensch.*, B, **60**, 153–9.

G. H. R. von Koenigswald 1957*b*, *Hemanthropus* N.g. not *Hemianthropus*, *Proc. Kon. Ned. Acad. Wetensch.*, B, **60**, 416.

3. _____

1. KIIK-KOBA, Crimea 1924.
2. **Homo kiik-kobiensis**, Bontch-Osmolovskii 1941.
   G. A. Bontch-Osmolovskii 1941, *Paleolit. Kryma.*, Vypusk 2, Moscow.
3. Holotype: hand bones, in Bontch-Osmolovskii 1941.

1. MODJOKERTO, Java 1936.
2. **Homo modjokertensis**, von Koenigswald 1936.
   G. H. R. von Koenigswald 1936, Erste Mitteilung über einen fossilen Hominiden aus dem Altpleistocän Ostjavas, *Proc. Kon. Ned. Acad. Wetensch. Amsterdam*, **39**, 1000–9.
3. Holotype: calvaria of child 1–3 years, in G. von Koenigswald 1936.
4. G. H. R. von Koenigswald 1940, Neue Pithecanthropus-Funde, 1936–1938: Ein Beitrag zur Kenntnis der Praehominiden, *Wet. Med. Dienst. Mijnb. Ned-Ind.*, Batavia, no. 28, 1–232.

1. MUGHARET EL-ZUTTIYEH, Israel 1925.
2. **Homo galilensis**, Joleaud 1931. Invalid **SN**.
   Homo primigenius galilaeensis, Hennig 1932. Invalid **SN**.
   (*Palaeanthropus palestinensis*, Weidenreich 1932.)[16]
   (*Palaeanthropus palestinus*, McCown & Keith 1934.)[17]
   (*Homo semiprimigenius palestinus*, Montandon 1943.)[18]
   L. Joleaud 1931, L'expansion géographique des Hommes de Néanderthal au Quaternaire Moyen dans les regions méditerranéennes, *Rev. Scientif*, **69**, 466–9.
   E. Hennig 1932, Fortschritte der Altsteinzeit-Forchung in der Alten Welt, *Petermann's Mitt.*, **78**, 134–7.
   F. Weidenreich 1932, Eine neuentdeckte Uebergansform zwischen dem Neandertaler und heutigen Menschen, *Nat. u. Mus.*, **62**, 384–9.
   T. D. McCown & A. Keith 1934, Palaeanthropus palestinus, *Proc. 1st. Int. Congr. Prehist. Protohist. Soc.* (London 1932), London.
   G. Montandon 1943, *L'homme préhistorique et les Préhumains*, Paris, 210–15.

1. MUGHARET ES-SKHUL, Israel 1931–2.
2. **Palaeanthropus palestinus**, McCown & Keith 1932.[19]
   Palaeanthropus palestinensis, Weidenreich 1932.[20]
   Homo semiprimigenius palestinus, Montandon 1943.[21]
   T. D. McCown & A. Keith 1934, Palaeanthropus palestinus,
   Proc. 1st Int. Congr. Prehist. Protohist. Soc. (London
   1932), London.
   F. Weidenreich 1932, Eine neuentdeckte Uebergansform
   zwischen dem Neandertaler und heutigen Menschen,
   Nat. u. Mus., **62**, 384–9.
   G. Montandon 1943, *L'Homme préhistorique et les
   Préhumains*, Paris, 210–15.
3. Lectotype: Child (Skhul I), in T. D. McCown & A. Keith
   1939, *The Stone Age of Mount Carmel: The Fossil
   Human Remains from the Levalloiso-Mousterian* (vol.
   II), Oxford.

1. MUGHARET ET-TABUN, Israel 1932–3.
2. (*Homo semiprimigenius palestinus*, Montandon 1943.)[22]
   Protanthropus tabunensis, Bonarelli 1944. Invalid **SN**.
   G. Montandon 1943, *L'homme préhistorique et les
   Préhumains*, Paris, 210–15.
   G. Bonarelli 1944, Sylloge Synonymica hominidarum
   fossilium hucusque cognitorum systematice ordinata,
   Ultima Miscellanea, **1**, 4, 1–67.
3. _____
4. T. McCown & A. Keith 1939, *The Stone Age of Mount
   Carmel: The Fossil Human Remains from the
   Levalloiso-Mousterian* (vol. II), Oxford.

1. NGANDONG, Java 1931.
2. **Homo (Javanthropus) soloensis**, Oppenoorth 1932.
   Homo primigenius asiaticus, Weidenreich 1932. **SH**.
   W. F. F. Oppenoorth 1932, *Homo (Javanthropus)
   soloensis*. Ein plistocene mensch van Java, *Wet. Meded.
   Dienst. Mijnb. Ned.-Ind.*, no. 20, 49–75.
   F. Weidenreich 1932, *Über pithekoide Meckmale bei
   Sinanthropus pekinensis und seine stammesge-
   schichtliche Beurteilung, Z. Anat. Entwicklungsges.*,
   99, 212–53.
3. Lectotype: Ngandong I calavaria, in Oppenoorth 1932.
4. F. Weidenreich 1951, Morphology of Solo Man, *Anthrop.
   pap. Amer. Mus.*, **43**, 205–90.

1. NISHIYAGI, Japan 1931.
2. **Nipponanthropus akasiensis**, Hasebe 1948.
   K. Hasebe, 1948, A Human coxal bone from lower
   Pleistocene deposits at Nishiyagi, *Zinruigaku Zassi (J.
   Anthrop. Soc. Nippon)*, **60**, 32–6.
3. Holotype: Os innominatum, in K. Hasebe 1948.
4. _____

1. SANGIRAN, Java 1939.
2. **Pithecanthropus dubius**, von Konigswald 1949.
   (*Meganthropus palaeojavanicus* von Koenigswald.) 1945.
   (In Weidenreich 1945.)[23]
   G. von Koenigswald 1949, The Fossil Hominids of Java, in
   R. W. van Bemmelen's *The Geology of Indonesia*, vol.
   I (in 2 vols.), The Hague, 106–11.
   F. Weidenreich 1945, Giant Early Man from Java and South
   China, *Anthrop. pap. Amer. Mus.*, **40**, 1–134. (A
   summary of this paper appeared earlier: *Science*, 16
   June 1944. The new species were mentioned.)
3. Holotype: Mandible fragment ('1939'), in F. Weidenreich
   1945.

1. SANGIRAN, Java 1939.
2. **Pithecanthropus robustus**, Weidenreich 1945.
   F. Weidenreich 1945, Giant Early Man from Java and South
   China, *Anthrop. pap. Amer. Mus.*, **40**, 1–134. (A summary of this paper appeared earlier: *Science,* 16 June
   1944. The new species were mentioned.)
3. Holotype: calvaria '*Pithecanthropus IV*,' in F. Weidenreich
   1945.

1. SANGIRAN, Java 1941.
2. **Meganthropus palaeojavanicus**, von Koenigswald (in
   Weidenreich 1945) 1950.
   F. Weidenreich 1945, Giant Early Man from Java and South
   China, *Anthrop. pap. Amer. Mus.*, **40**, 1–134. (A summary of this paper appeared earlier: *Science*, 16 June
   1944. The new species were mentioned.)
   G. von Koenigswald 1950, Fossils Hominids from the Lower
   Pleistocene of Java, *Rep. 18th Internat. Geol. Congr.
   1948*, London, part 9, 59–61.
3. Lectotype: mandible fragment ('1941'), in F. Weidenreich
   1945.

1. SHANIDAR, Iraq 1953.
2. **Homo sapiens shanidarensis**, Senyürek 1957.
   M. Senyürek 1957, The Skeleton of the fossil infant found in Shanidar, *Anatolia*, **2**, 49–55.
3. Holotype: teeth of infant of 1 year, in Senyürek 1957.

1. TRINIL, Java 1891.
2. **Anthropopithecus erectus**, Dubois 1892.[24]
   Pithecanthropus, Anon. 1893, Dubois 1894.
   Homo javanensis primigenius, Houzé 1896.
   Homo pithecanthropus, Manouvrier 1896.
   Hylobates giganteus, Bumuller 1899.
   Proanthropus, Wilser 1900.
   Pithecanthropus duboisii, Morselli 1901.
   Hylobates gigas, Krause 1909.
   Homo trinilis, Alsberg 1922.
   Praehomo asiaticus javanensis, Eickstedt 1932.
   E. Dubois 1892, Bivoegsel tot de Javasche Conrant, *Verslag. Mijnw.*, Batavia, 3rd quarter, 10–14.
   Anonymous 1893, *Verslag Mijnw.*, Batavia, 4th quarter.
   E. Dubois 1894, Pithecanthropus erectus. Eine menschenähnliche Ubergangsform aus Java, Batavia. (Reprinted in *Jaarboek van het Mijnwezen*, **24**, 5–77, 1895.)
   E. Houzé 1896, *Rev. Univ. Bruxelles*, **1**.
   L. Manouvrier 1896, Réponse aux objections contre le Pithecanthropus, *Bull. Soc. Anthrop. Paris*, **7**, 396–460.
   J. Bumuller 1899, *Das menschliche Femur nebst Beitragen zur Kenntnis der Affen-femora*, **12**, *Das Femur des Pithecanthropus erectus*, Ausberg, 124–38.
   L. Wilser 1900, Der Pithecanthropus erectus und die Abstammung des Menschen. *Verh. Naturw. Ver. Karlsruhe*, **13**, 551–76.
   E. Morselli 1901, *Il precursore dell'uomo. Nota riassuntiva*, Genoa, 1–19.
   W. Krause 1909, Anatomie der Menschenrassen, in Bardeleben, *Handbuch der Anatomie des Menschen*, **1**, (3), Jena, 173.
   P. Alsberg 1922, *Das Menscheitsratsel*, Dresden, 398.
   E. Eickstedt 1932, Hominiden und Simioiden Über den derzeitigen Stand der Abstammungsfrage, *Z. arztl. Fortbild.*, **29**, 608–13.
3. Lectotype: calotte ('Pithecanthropus I'), in Dubois 1894.[25]

1. WADJAK, Java 1889–90.
2. **Homo wadjakensis**, Dubois 1921.
   E. Dubois 1921, The Proto-Australian Fossil Man of Wadjak, Java, *Proc. Kon. Ned. Acad. Wetensch. Amsterdam*, **23**, 1013–51.
3. Lectotype: cranium of adult female ('Wadjak I'), in E. Dubois 1921.

Europe

1. LA CHAPELLE-AUX-SAINTS, France 1908.
2. **Homo chapellensis**, Buttel-Reepen 1911.
   Archanthropus primigenius, Abel 1920. **GH**.
   H. von Buttel-Reepen 1911, Der Urmensch vor und Während der Eiszeit in Europe Ein Sammelreferat. *Naturw. Woecherschr.*, **26**, 177–89, 193–204, 209–19, 225–31.
   O. Abel 1920, Hominiden, in *Lehrbuch der Palaozoologie*, **1**, 456–7, Jena.
3. Holotype: adult male skeleton, in Boule 1911–13, L'Homme fossile de la Chapelle-aux-Saints, *Ann. Paleont.*, **6**, 106–72; **7**, 21–192; **8**, 1–70.

1. CHANCELADE, France 1888.
2. **Homo priscus**, Lapouge 1899.[26]
   (*Notanthropus eurafricanus recens* Sergi, 1911.)[27]
   G. V. de Lapouge 1899, *L'aryen: son rôle sociale*, Paris.
   G. Sergi 1911, *L'uomo secondo le origine, l'antichita, le variazioni e la distribuzione geographica*, Turin.
3. Lectotype: adult male skeleton of 35–40 years, in H. V. Vallois 1946, Nouvelles Recherches sur le Squelette de Chancelade, *L'Anthropologie, Paris*, **50**, 166–202.

1. COMBE CAPELLE, France 1909.
2. **Homo aurignacensis hauseri**,[28] Klaatsch & Hauser 1910.
   Notanthropus eurafricanus recens, Sergi 1911.[29]
   Homo fossilis proto-aethiopicus, Giuffrida-Ruggeri 1915.[30]
   Homo meridionalis proto-aethiopicus, Giuffrida-Ruggeri 1921.[31]
   (*Homo sapiens cro-magnonensis*, Gregory 1921.)[32]
   H. Klaatsch & O. Hauser 1910, *Homo aurignacensis Hauseri*, ein palaeolithischer Skelettfund aus dem unteren Aurignacien der Station Combecapelle bei Montferrand (Perigord), *Prahist Z.*. **1**, 273–338.

G. Sergi 1911, *L'uomo secondo le origine, l'antichita, le variazioni e la distribuzione geografica*, Turin.

V. Giuffrida-Ruggeri 1915, Quatro crani-prehistorici dell'Italia meridionale, *Arch. Antrop. Etnol.*, **45**, 292–315.

V. Giuffrida-Ruggeri 1921, *Su l'origine del uomo. Neove teorie e documenti*, Bologna.

W. K. Gregory 1921, *The Origin and Evolution of the Human Dentition*, Baltimore.

3. Holotype: adult male skeleton 40–50 years, in Klaatsch & Hauser 1910.

1. LA COTTE DE SAINT BRELADE, Jersey, England 1910-11.
2. **Homo breladensis**, Marett 1911.
   R. R. Marett 1911, Pleistocene Man in Jersey, *Archaeologia*, **62**, 449–80.
3. Holotype: 9 teeth of adult, in Marett 1911.
4. A. Keith & F. H. S. Knowles 1911, A Description of Teeth of Paleolithic Man from Jersey, *J. Anat. Physiol.*, **46**, 12–27.

1. CRO-MAGNON, France 1868.
2. **Homo spelaeus**, Lapouge 1899.[33]
   Homo sapiens cro-magnonensis, Gregory 1921.[34]
   Homo laterti, Pycraft 1925. Invalid **SM**.
   G. V. de Lapouge 1899, *L'Aryen: son rôle sociale*, Paris.
   W. K. Gregory 1921, *The Origin and Evolution of the Human Dentition*, Baltimore.
   W. P. Pycraft 1925, Diagnoses of four Species and one Subspecies of the Genus *Homo*, *Man*, **25**, 162–4.
3. Lectotype: cranium of adult male (No. 1), in Broca 1868; in L. Lartet & H. Christy 1865–75. *Reliquiae Aquitanicae*, London, 97–122.
4. P. Broca 1868, Sur les cranes et ossements des Eyzies, *Bull. Soc. Anthrop. Paris*, **3**, 350–92, 416–46, 554–74.

1. EGUISHEIM Alsace, France 1865.
2. (*Notanthropus eurafricanus archaius*, Sergi 1911.)[35]
   G. Sergi 1911, *L'uomo secondo le origine, l'antichita, le variazioni e la distribuzione geografica*, Turin.
3. _____
4. G. Schwalbe 1902, Der Schädel von Egisheim, *Beitrage z. Anthrop. Elsaaz-Lothringens*, **3**, 1–64.

1. GIBRALTAR (Forbes Quarry), 1848.
2. **Homo calpicus**, Keith 1911.[36]
   Homo gibraltarensis, Battaglia 1924.
   A. Keith 1911, The Early History of the Gibraltar Cranium, *Nature Lond.*, **87**, 313–14.
   R. Battaglia 1924, Osservazioni su l'uomo fossile di Broken Hill, *Boll. Soc. Adriatica Sci. Nat.*, **28**, (2), 305–22.
3. Holotype: adult calvaria (Gibraltar I), in W. J. Sollas 1908, On the Cranial and Facial Characters of the Neanderthal Race, *Phil. Trans. Roy. Soc. London*, **199**, B, 281–339.

1. GRIMALDI, (Cave III, Barma Grande) France 1884–92–94.
2. (*Homo spelaeus*, Lapouge 1899.)[37]
   (*Notanthropus eurafricanus recens*, Sergi 1911.)[38]
   G. V. de Lapouge 1899, *L'Aryen: son rôle sociale*, Paris.
   G. Sergi 1911, *L'uomo secondo le origine, l'antichita, le variazioni e la distribuzione geografica*, Turin.
3. _____
4. R. Verneau 1906, in L. de Villeneuve et al. 1906–12, *Les Grottes de Grimaldi*, Monaco.

1. GRIMALDI, (Cave I, Grotte des Enfants) France 1874–5, 1901.
2. **Homo grimaldii**, Lapouge 1905–6.
   (*Homo spelaeus*, Lapouge 1899.)[39]
   Homo niger, Wilser 1903. **SH**.
   (*Notanthropus eurafricanus recens*, Sergi 1911.)[40]
   (*Homo sapiens grimaldiensis*, Gregory 1921.)[41]
   Homo grimaldicus, Hilber 1922. Invalid **SM**.
   G. V. de Lapouge 1905–6, Die Rassengeschichte der französischen Nation, *Polit. Anthrop. Rev.*, **4**, 16–24.
   G. V. de Lapouge 1899, *L'Aryen: son rôle sociale*, Paris.
   L. Wilser 1903, Die Namen der Menschenrassen, *Globus*, **84**, 303–7.
   G. Sergi 1911, *L'uomo secondo le origine, l'antichita, le variazioni e la distribuzione geografica*, Turin.
   W. K. Gregory 1921, *The Origin and Evolution of the Human Dentition*, Baltimore.
   V. Hilber 1922, Urgeschichte Steiermarks, *Mitt. Nat. Var. Steiermark*, **58**, (B), 1–11.
3. Lectotype: young male ('Negroid') from 7.75 m. level in R. Verneau 1906. *Les Grottes de Grimaldi, Anthropologie*, vol. I, Monaco.

1. KRAPINA, Jugoslavia 1899–1905.
2. **Homo neanderthalensis** var. **krapinensis** Gorjanovic-
   Kramberger 1902.[42]
   Homo antiquus, Adloff 1908.
   Homo alpinus, Krause 1909. **SH**.
   Palaeanthropus krapiniensis, Sergi 1911[43] **SH**.
   K. Gorjanovic-Kramberger 1902, Der palaeolitische Mensch
   und seine Zeitgenossen aus dem Diluvium von Krapina
   in Kroatien, *Mitt. Anthrop. Ges. Wien*, **32**, 189–216.
   P. Adloff 1908, Die Zahne des Homo primigenius von
   Krapina, *Anat. Anz.*, **32**, 301–2.
   W. Krause 1909, Anatomie der Menschenrassen, in
   Bardeleben, *Handbuch der Anatomie des Menschen*,
   1, 3, Jena.
   G. Sergi 1911, *L'uomo secondo le origine, l'antichita, le
   variazioni e la distribuzione geografica*, Turin.
3. Holotype: Krapina calvaria 'D,' in Gorjanovic-Kramberger
   1906, Der Diluviale mensch von Krapina in Kroatien,
   in: *Studien uber die Entwicklungsmechanik des
   Primatenskelettes*, Wiesbaden.

1. LAUGERIE-BASSE, France 1872.
2. (*Homo priscus*, Lapouge 1899.)[44]
   (*Notanthropus eurafricanus recens*, Sergi 1911.)[45]
   G. V. de Lapouge 1899, *L'Aryen: son rôle sociale*, Paris.
   G. Sergi 1911, *L'uomo secondo le origine, l'antichita, le
   variazioni e la distribuzione geografica*, Turin.
3. _____
4. E. T. Hamy 1874, Description d'un squelette humain fossile
   de Laugerie-Basse, *Bull. Soc. Anthrop. Paris*, **9**, 652–8.
   A. Quatrefages & E. T. Hamy 1882, Races Humaines
   Fossiles, in *Crania Ethnica*, Paris.

1. LEUCA, Italy 1961.
2. **Homo homo** var. **neanderthalensis**, Blanc[46] 1961. Invalid
   SN.
   A. C. Blanc 1961, Leuca I, Il primo reperto fossile
   neandertaliano del Salento, *Quaternaria*, **5**, 271–8.
3. Holotype: Left permanent M2, in Blanc 1961.

1. MAUER, Germany 1907.
2. **Homo heidelbergensis**, Schoetensack 1908.
   Palaeanthropus heidelbergensis (Schoet. 1908), Bonarelli
   1909.

Pseudhomo heidelbergensis (Schoet. 1908), Ameghino 1909.
Protanthropus heidelbergensis (Schoet. 1908), Arldt 1915.
Praehomo heidelbergensis (Schoet. 1908), Eikstedt 1932.
Praehomo europaeus, Eikstedt 1934.
Anthropus heidelbergensis (Schoet. 1908), Weinert 1937. **GH**.
Maueranthropus heidelbergensis (Schoet. 1908), Montandon 1943.
Europanthropus heidelbergensis (Schoet. 1908), Wüst 1950.
Euranthropus, Arambourg 1955. Invalid **GB**.
O. Schoetensack 1908, Der Unterkiefer des *Homo heidelbergensis* aus den Sanden von Mauer bei Heidelberg, *Ein Beitrag zur Palaontologie des Menschen*, Leipzig.
G. Bonarelli 1909, Palaeanthropus (n.g.) heidelbergensis (Schoet.), *Riv. Ital. Paleont.*, **15**, 26–31.
Fl. Ameghino 1909, Le Diprothomo platensis, un précurseur de L'homme du pliocene inférieur de Buenos Aires, *An. Mus. Nac. Buenos Aires*, **19**, 107–209.
T. Arldt 1915, Die Stammesgeschichte der Primaten und die Entwicklung der Menschenrassen, *Fortschr. Rassenk*, **1**, 1–52.
E. Eickstedt 1932, Hominiden und Simioiden. Uber den derzeitigen Stand der Abstammungsfrage, *Z. arztl. Fortbild.*, **29**, 608–13.
E. Eickstedt 1934, *Rassenkunde und Rassengeschichte der Menschheit*, Stuttgart.
H. Weinert 1937, Der Unterkiefer von Mauer zur 30 jahrigen Wiederkehr seiner Entdeckung, *Z. Morph. Anthrop.*, **37**, 102–13.
R. Montandon 1943, *L'homme préhistorique et les préhumains*, Paris.
K. Wüst, 1950, Uber den Unterkiefer von Mauer (Heidelberg) im Vergleich zu anderen fossilen und renzenten Unterkeifern von Anthropoiden und Hominiden mit desonderer Berucksichtigung der phyletischen Stellung des Heidelberger Fossils, *Z. Morph. Anthrop.*, **42**, 1–112.
C. Arambourg 1955, Problèmes actuels de Paléontologie, *Colloques Internat. Paris*, **60**, 135–148.
3. Holotype: mandible, in Schoetensack 1908.

1. LE MOUSTIER, France 1908.
2. **Homo transprimigenius mousteriensis**, Forrer 1908.[47]
   Homo mousteriensis hauseri, Klaatsch & Hauser 1909.[48]
   Homo acheulensis moustieri, Wiegers 1915.
   Homo le mousteriensis, Wiegers 1915. Invalid.[49]
   R. Forrer 1908, *Urgeschichte des Europäers*, Stuttgart, 1908.
   H. Klaatsch & O. Hauser 1909, *Homo mousteriensis Hauseri*, ein altdiluvialer Skelettfund im Department Dordogne unde sein Zugelhorigkeit zum Neandertaltypus, *Arch f. Anthrop.*, **35**, (N.F.7), 287–97.
   F. Wiegers 1915, Das geologische Alter des Homo Mousteriensis, *Z. f. Ethnol.*, **47**, 68–72.
3. Holotype: skeleton of 18 year old male, in Klaatsch & Hauser 1909. (Destroyed 1944).
4. H. Weinert 1925, *Der Schadel des eiszeitlichen Menschen von Le Moustier in neuer Zusammensetzung*, Berlin, 1925.

1. LA NAULETTE, Belgium 1866.
2. **Homo naulettensis**, Baudouin 1916. Invalid **SM**.
   M. Baudouin 1916, Sur L'anteriorité de la machoire trouvée a la Naulette, *C.R. Acad. Sci.*, **162**, 519–20.
3. _____
4. E. Dupont 1866, Étude sur les fouilles scientifiques exécutées pendant l'hiver de 1865–1866 dans les cavernes des bordes de la Lesse, *Bull. Acad. Roy. Belgique*, cl. des Sciences, **22**, 47–52.
   P. Topinard 1886, Les caractères simian de la machoire de La Naulette, *Rev. d'Anthrop.*, **1**, 385–431.

1. NEANDERTAL, Germany 1856.
2. **Homo neanderthalensis**, King 1864.
   Homo (Protanthropus) neanderthalensis (King), Bonarelli 1909.
   Protanthropus atavus, Haeckel 1895. Invalid **SM**.
   Homo europaeus primigenius, Wilser 1898. Invalid **SM**.
   Homo primigenius, Schwalbe 1903.
   Palaeanthropus europaeus, Sergi 1910.
   Archanthropus, Arldt 1915.
   Anthropus neanderthalensis (King 1864), Boyd-Dawkins 1926.
   Metanthropus, Sollas 1933. Invalid **GB**.
   W. B. R. King 1864, The Reputed Fossil Man of the Neanderthal, *Quart. J. Sci.*, **1**, 88–97.

G. Bonarelli 1909. Le razze umane e le loro probabile affinita. *Boll. Soc. Geogr. Ital.* Fasc. 8:827–51, 9:953–79.

E. Haeckel 1895, *Systematische Phylogenie der Wirbeltiere*, III, Berlin.

L. Wilser 1898, Menschenrassen und Welgeschichte, *Naturw. Wochenschr.*, **13**, 1–8.

G. Schwalbe 1903, *Die Vorgeschichte des Menschen*, Braunschweig.

G. Sergi 1910, Paléontologie Sud-américaine, *Scientia* **8**, 465–75.

T. Arldt 1915, Die Stammesgeschichte der Primaten und die Entwicklung der Menschenrassen, *Fortschr. Rassenk.*, , **1**–52.

W. Boyd-Dawkins 1926, The Range of the Anthropus Neanderthalensis on the Pleistocene Continent, *Rep. Brit. Ass.*, **94**, 386.

W. J. Sollas 1933, The Sagittal Section of the Human Skull, *J. R. Anthrop. Inst.*, **63**, 389–431.

3 . Holotype: cranium and skeletal remains of adult of 45–50 years, in H. Schaafhausen 1888, *Der Neandertaler Fund*, Bonn.

1. OBERKASSEL, Germany 1914.
2 . **Homo mediterraneus fossilis**, Behm 1915. **SH**.
   (*Homo sapiens cro-magnonensis*, Gregory 1921.)[50]
   H. W. Behm 1915, Die Fossilmenschenfunde von Oldoway und Oberkassel, *Prometheus*, **26**, 161–4.
   W. K. Gregory 1921, *The Origin and Evolution of the Human Dentition*, Baltimore.
3 . _____
4 . M. Verworn et al. 1919, *Der diluviale Menschenfund von Obercassel bei Bonn*, Wiesbaden.

1. PREDMOST, Czechoslovakia 1884–1928.
2 . **Notanthropus eurafricanus archaius**, Sergi 1911.[51]
   Homo predmostensis, Absolon 1920.
   Homo predmosti, Matiegka 1938.
   G. Sergi 1911, *L'uomo secondo le origine, l'antichita, le variazioni e la distribuzione geographica*, Turin.
   K. Absolon 1920, Die Funde von Predmost, in H. Klaatsch & A. Heilborn 1920, *Der Werdegang der Menschheit und die Entstehung der Kultur*, Berlin, 357–73.
   J. Matiegka 1938, *Homo Predmostensis fosilni clovek z Predmosti na Morave. II; Ostatni casti kostrove*, Prague.

3. Lectotype: specimen No. XXI, in Matiegka 1938.

1. SACCOPASTORE, Italy 1929, 1935.
2. **Homo neanderthalensis** var. **aniensis**, Sergi 1935.[52]
   S. Sergi 1935, Die Entdeckung eines weiteren Schädels des *Homo neandertalensis* var. *aniensis*, in der Grube von Saccopastore (Rom), *Anthrop. Anzeiger*, **12**, 281–4.
3. Lectotype: Calvaria of adult female ('Saccopastore I'), in S. Sergi 1944, Craniometriae craniografia del primo paleantropo di Saccopastore, *Richerche di Morph.*, Roma, **20** & **21**, 733–91.

1. SPY, Belgium 1886.
2. **Homo spyensis**, Krause 1909.
   Homo priscus, Krause 1909.[53] **SH**.
   W. Krause 1909, Anatomie der Menschenrassen, in Bardeleben, *Handbuch der Anatomie des Menschen*, **1**, 3, Jena.
3. Lectotype: Adult male? skeleton of 35 years ('Spy I'), in J. Fraipont & M. Lohest 1887, La Race humaine de Néanderthal ou de Canstadt en Belgique, *Arch. Biol.*, **7**, 587–757.

1. STEINHEIM, Germany 1933.
2. **Homo steinheimensis**, Berckhemer 1936.
   Homo murrensis, Weinert 1936.[54]
   F. Berckhemer 1936, Der Urmenschenschädel aus den zwischeneiszeitlichen Fluss-Schottern von Steinheim an der Murr, *Forsch. Fortsch.*, **12**, 349–50.
   H. Weinert 1936, Der Urmenschenschädel von Steinheim, *Z. Morph. Anthrop.*, **35**, 463–518.
3. Holotype: adult female calvaria, in Berckhemer 1936.
4. H. Weinert 1936.

1. SWANSCOMBE, Great Britain 1935 & 1955.
2. **Homo sapiens proto-sapiens**, Montandon 1943.[55]
   Homo marstoni, Paterson 1940. Invalid **SN**.
   Homo swanscombensis, Kennard 1942. Invalid **SN**.
   G. Montandon 1943, *L'homme préhistorique et les préhumains*, Paris.
   T. T. Paterson 1940, Geology and Early Man, *Nature Lond.*, **146**, 12–15, 49–52.
   A. S. Kennard 1942, Faunas of the High terrace at Swanscombe, *Proc. Geol. Assoc.*, **53**, 105.

3. Lectotype: occipital and left parietal bones, in W. Le G. Clark 1938, Report on the Swanscombe Skull, *J. R. Anthrop. Inst.*, **68**, 58–97.
4. C. D. Ovey 1964, *The Swanscombe Skull*, London.

1. WEIMAR-EHRINGSDORF (Kaempfer's Quarry), Germany 1914–16.
2. **Homo heringsdorfensis**, Moller 1928. Invalid **SM**.
   Homo ehringsdorfensis, Paterson 1940. Invalid **SN**.
   A. Moller 1928, see E. Werth 1928, *Der fossile Mensch. Grundzüge einer Paläanthropologie*, Berlin, 201.
   T. T. Paterson 1940, Geology and Early Man, *Nature Lond.*, **146**, 12–15, 49–52.
3. _____
4. H. Virchow 1920, *Die Menschlichen Skeletreste aus dem Kampfe'schen Bruch im Travertin von Ehringsdorf bei Weimar*, Jena.

3. Assessment of the State of Hominid Nomenclature

My examination of the literature relating to fossils which are believed to be of Pleistocene age has yielded the information that 60 different specimens have been made the types of new Hominid taxa. These named taxa include 19 new genera and 55 new species and subspecies which are valid and have priority for the taxa for which they were created. Twelve further generic and 32 specific names are objective junior synonyms of these named taxa. Beyond this there are 4 generic and 17 specific names which are invalid since they do not comply with the code of nomenclature, as explained in the Introduction. Attempts to create new taxa also include 2 generic and 7 specific homonyms.

The position, then, is such that we have separate named species for almost every fossil find of any importance from the Pleistocene, a state of affairs which does not in any way reflect the actual taxonomy of evolving man. It is not so much that early workers were 'splitters,' in the taxonomic sense, but that they were ignorant of the meaning of the concept of the species, and used binomial nomenclature as a system of labelling. This publication is prepared as a guide in the current movement which aims at a revision of the classification of the Hominidae so that its structure reflects as nearly as possible the reality of evolving populations as they replaced each other during the Pleistocene.

In view of the wealth of type specimens from all geological levels of the Pleistocene, it would appear extremely improbable that any further discoveries could warrant the creation of a new taxon except at the subspecific level. Any new taxa which are created in future on the basis of Pleistocene fossils could only be justified by very careful and extensive comparative examination with existing fossil specimens. New finds would have to be shown to fall well clear of the range of variability of existing taxa, and that range should be computed by comparison with living species and subspecies, *not* on the basis of the variability of a limited number of incomplete fossil specimens. These comments would have been equally appropriate in 1955, yet since the beginning of that year we have seen the creation of ten new Hominid taxa including two genera. Time will show that none of these genera is justified, and only a few of the specific names will survive as subspecies.

It is to be hoped that as the principles underlying classification and nomenclature are better understood, a certain restraint will be apparent in the creation of new latin names. It is essential to remember that the recognition of a new Hominid species subsumes a hypothesis about human evolution, and hypotheses should not be published unless the facts available at the time of their publication make them effectively demonstrable.[56]

The Art of Taxonomic Revision

Anyone is at liberty to express his views of the phyletic relationship of the fossil Hominidae by publishing a revised classification. New classifications are published from time to time,[57] and they usually involve a great simplification of the position represented by the names listed above. However, these names will not necessarily be forgotten, since, just as modern man has six recognized subspecies, so any fossil species, though with a more restricted geographical range, may be reasonably expected to be (ultimately) divisible into just as many subspecies at any moment in time, for one can assume a slower rate of gene flow in antiquity. If, furthermore, the fact is taken into account that a palaeospecies has a dimension in time equivalent to that in space (that is, a dimension of morphological variation which is equivalent), then we can expect a palaeospecies to be represented by 6 x 6 subspecies. We should not be surprised, therefore, to discover as many as 30–40 subspecies in a palaeospecies with a distribution throughout the Old World, such as *Homo erectus* (as defined today). Many of the specific names, listed here, may therefore be preserved as palaeontological subspecies.

An example of taxonomic revision follows. Let us suppose that it is desired to make the fossils from a number of sites the types of a range of subspecies of one species. In the left-hand column of the following table the sites that I have selected are listed in alphabetical order. In order to decide on their specific name, we take the oldest which is valid from among them, in this instance that given to the Trinil holotype—*Pithecanthropus erectus*.[58] It is now generally agreed that this species, *P. erectus*, should be transferred to the genus *Homo*, and this we are at liberty to do since we believe that the morphological difference between them is insufficient to justify generic distinction.[59]

All the sites in our list will now become sub-species of *Homo erectus*. Their specific names can be used as subspecific names, and added to this binomial to form a trinomial. Anyone who is familiar with the fossils mentioned will observe that some of our subspecies are approximately contemporary, while others are certainly sequent in time. They can therefore be arranged in a three-dimensional sequence, which can be summarized in the two dimensions of Table 2, according to the grade of their evolutionary progress towards *Homo sapiens*, and their chronological position.[60]

This approach to taxonomic revision not only brings Hominid taxonomy into line with that of other vertebrate families, but also enables us to retain the species-group names which have been historically associated with the fossils. The author of the names could also be retained in a formal classification, and placed in brackets. Thus historical continuity will be preserved. Table 2 also suggests where we might expect to find further subspecies both in space and time. In particular we might expect a fourth grade (and even a fifth) linking these subspecies of *Homo erectus* with the early

Table 1

Site	Valid Name	Revised Classification
Choukoutien	Sinanthropus pekinensis	Homo erectus pekinensis
Mauer	Palaeanthropus heidelbergensis	Homo erectus heidelbergensis
Olduvai (Upper Bed II)	Homo leakeyi	Homo erectus leakeyi
Olduvai (Lower Bed II)	Homo habilis*	Homo erectus habilis*
Ternifine	Atlanthropus mauritanicus	Homo erectus mauritanicus
Trinil	Pithecanthropus erectus	Homo erectus erectus
Modjokerto	Homo modjokertensis	Homo erectus modjokertensis
Swartkrans (1949)	Telanthropus capensis	Homo erectus capensis**

* The holotype of *Homo habilis* is in fact the rather more primitive Bed I specimen. If this classification were seriously adopted, a new subspecific name would have to be given to the Bed II specimen.
** The species-group name *capensis* has already been used in this genus for the Boskop skull. If this classification were seriously adopted, a new name would have to be given to this subspecies.

Table 2

Grade	Subspecies of *Homo erectus* in Space and Time					
	Europe	N. Africa	E. Africa	S. Africa	E. Asia	S.E. Asia
3	heidel-bergensis	mauritan-icus	leakeyi		pekinensis	
2						erectus
1			habilis	capensis		modjoker-tensis

direction of evolution ↑

and surprisingly modern *Homo sapiens* fossils from Steinheim and Swanscombe.

My intention in this monograph has been to show how the classification of the Hominidae might now be revised on the basis of a historically correct nomenclature which has been carefully examined according to the International Code. It is to be hoped that whatever classification is adopted in the future, it will be clear what is the proper nomenclature for any new taxon created on the basis of existing specimens, and that therefore nomenclatural confusion will be reduced.

Nomenclature may not appear at first sight to be of central importance in Biology, but its importance is fundamental to successful communication. Confucius was in no doubt about this, for when he was asked what was to be his first action on taking over the government of a state he is reported to have replied:[61]

> If I must begin, I would begin by defining the names of things. If names of things are not properly defined, words will not correspond to facts. When words do not correspond to facts, it is not possible to perfect anything. Therefore a wise man will always specify what he names; will always be exact in the words he uses.

Endnotes

[1] For a full discussion of these problems, see my article, Science and Human Evolution, *Nature*, 203, 448–51, 1 August 1964.

[2] A term proposed by Simpson (singular *nomen*) for the binomial latin names given to zoological species under the rules of zoological nomenclature. See also the reference given in endnote 1, above.

[3] C. Linnaeus, 1758, *Systema Naturae*, 10th edn. Stockholm.

[4] B. G. Bory de St. Vincent, 1825, *Dict. Class. Hist. Nat.*, 8, 269, Paris.

[5] *Code International de Nomenclature Zoologique*, International Trust for Zoological Nomenclature, London, 1961. Hereinafter referred to as CINZ.

[6] See CINZ, Art. 73.

[7] See CINZ, Art. 74.

[8] Where a fossil has lost its status as a type, the name previously given to it is placed in brackets.

[9] *Boskop* is not a latinized name and does therefore not conform to CINZ, Article 11. (See also Note 4).

[10] Also published in error as *Scyphanthropus*—erroneous subsequent spelling. (See Note 3).

[11] Genus here considered to be *Palaeanthropus* Bonarelli 1909—see Note 3.

[12] The Holotype is *not* the skull (S.1) later selected as 'Type' (presumably intended to be lectotype) in R. Broom and G. Schepers 1946, The South African Fossil Ape-Men—The *Australopithecinae, Trans. Mus. Mem.*, **2**, 1–272.

[13] This name is considered to be an erroneous subsequent spelling (see Note 3): *pekingensis* does not therefore rank as an objective junior synonym.

[14] These fossils listed amongst others as syntypes of this taxon; they here lose their status as types. For lectotype see Mugharet Es-Skhul.

[15] The genus *Hemianthropus* was introduced by Freudenberg in 1929 for the skeletal fragments from Bammental (near Mauer) known as Bammental II. These are now considered to be *Homo sapiens*, so that *Hemianthropus* is an objective junior synonym of *Homo*. As a result of this the author changed the name to *Hemanthropus* later the same year. Such a change is allowed by CINZ but the original genus was not accompanied by a statement purporting to give characters differentiating the taxon, and is therefore a nomen nudum.

[16] This specimen is cited as one of many syntypes of these taxa. It here loses its status as a type. For the lectotype see Mugharet Es-Skhul. See note under Mugharet Es-Skhul which refers to this name.

[17] This specimen is cited as one of many syntypes of these taxa. It here loses its status as a type. For the lectotype see Mugharet Es-Skhul.

[18] Ibid.

[19] Specimens from this site and Mugharet El-Zuttiyeh are treated as syntypes by these authors. No. I from this site is here selected as the lectotype.

[20] The name *Palaeanthropus palestinus* was first used by McCown and Keith to describe the Galilee skull and the Skhul child (Skhul I) in a paper orally presented at Cambridge in 1932 and published in 1934. In this paper the special characters of the Skhul child and the Galilee skull were indicated, and the paper constituted a valid description for the purposes of Article 13. The adult skeletons from the Skhul cave were only mentioned. Weidenreich, who was at the conference, published a report the same year (1932) in which he attributed the name *Palaeanthropus palestinensis* to McCown and Keith. It seems best to consider this spelling to be an incorrect original spelling and to consider McCown and Keith's spelling of 1934 and later publications, a valid emendation. (See Note 3.) According to the Rules, the valid name of this taxon is *Palaeanthropus palestinus* McCown and Keith 1932 (Articles 32 and 33).

[21] Specimens from this site and Mugharet el-Zuttiyeh, et-Tabun and Djebel Kafzeh are listed as syntypes of this taxon. Skhul I is here created the lectotype.

[22] These fossils listed amongst others as syntypes of this taxon. They here lose their status as types. For lectotype see Mugharet Es-Skhul.

[23] The mandible here loses its status as a syntype of this taxon. For lectotype see Sangiran 1941 mandible.

[24] This generic name was introduced by de Blainville in 1838 for the Chimpanzee (*A. troglodytes*). It is an objective junior synonym of *Pan* Oken, 1816.

[25] New family *Pithecanthropidae* created in this publication ('between Simiidae and Hominidae').

[26] Specimen published as a syntype with Laugerie-Basse (q.v.); it is here made a lectotype.

[27] Specimen listed as a syntype with Combe Capelle (q.v.); it here loses this status.

[28] Not clear whether intended as subspecific or author's name; if subspecific, Le Moustier fossil has priority (1909).

[29] This fossil is listed by Sergi (1911) as one of seven syntypes. It is here created a lectotype. (See also Chancelade & Laugerie-Basse.)

[30] Associated with *H. aurignacensis* as an alternative. (See Note 6.)

[31] Subspecies *proto-aethiopicus* transferred to a new species by V. Giuffrida-Ruggeri.

[32] Specimen listed by Gregory as a syntype. It here loses this status. (See Cro-Magnon.)

[33] These fossils were listed amongst others as syntypes of this taxon. The adult male is here created the lectotype. (See also Grimaldi.)

[34] These fossils were quoted amongst others as syntypes of this taxon. The adult male is here created a lectotype. (See also Oberkassel and Combe Capelle.)

[35] Specimen listed as a syntype with Combe Capelle (q.v.); it here loses this status.

[36] This name was proposed by Falconer in 1864, but CINZ, Article 50, makes it clear that a name is to be attributed to the author by whom it was first published under conditions which satisfied the requirements of Articles 7–20.

[37] These remains included as syntypes of this taxon: they here lose their status as types. (For lectotype see Cro-Magnon.)

[38] These skeletal remains are listed by Sergi 1911, as syntypes of this taxon: they here lose this status. (For lectotype see Combe Capelle.)

[39] Two children from 2.70 metre level (discovered 1874–5) listed as syntypes of this taxon: they here lose their status as types (see Cro-Magnon).

[40] Male adult specimen from 7.05 metre level (discovered 1901) listed by Sergi as a syntype of this taxon: it here loses this status (see Combe Capelle).

[41] Male and female skeletons from 7.75 metre level listed as syntypes of this subspecies by Gregory 1921.

[42] Name of variety is available. (See Note 5.) Holotype taken to be the parietal and occipital fragments described in conjunction with this name by Gorjanovic-Kramberger in 1902 and considered here (as by the author) to belong to a single brachycephalic individual. They were later known as skull 'D.' (See Gorjanovic-Kramberger 1906, p. 107.)

[43] G. Sergi 1911, quotes Krapina 'A' (Gorjanovic-Kramberger 1906, p. 89) as holotype of this taxon, and separates Krapina B & C which he places in *Palaeanthropus europaeus*, Sergi 1910. (See Neandertal.) But the name *Krapiniensis* is not available, since the difference in spelling is not considered to constitute the creation of a new name. (See Note 3.) In fact Sergi first used this binomen in 1910, but without any indication of the fossil to which it applied. G. Sergi 1910, Paléontologie sud-américaine, *Scientia*, **8**, 465–75.

[44] Specimen listed as a syntype: it here loses this status (see Chancelade).

[45] Specimen listed as a syntype: it here loses this status (see Combe Capelle).

[46] Proposed as a synonym of *H. sapiens* var. *neanderthalensis*.

[47] *transprimigenius* is the specific name, and has priority over *mousteriensis* where a specific distinction is required.

[48] Not clear whether intended as subspecific name or author's name. Presumed subspecific.

[49] Under CINZ, Article 11, a single word is required to designate the species.

[50] Specimen listed as a syntype: it here loses this status (see Cro-Magnon).

[51] All specimens included as syntypes of this subspecies. Here specimen No. XXI is selected as a lectotype of the taxon, being the most representative adult individual remaining extant (see also Eguisheim). The majority of the syntypes were destroyed in 1944.

[52] Both calvaria listed as syntypes of this variety: here 1929 skull is created lectotype of this taxon. Name of variety is valid. (See Note 5.)

[53] Given as an alternative to *spyensis*.

[54] Weinert suggests this name as an alternative to *H. Steinheimensis*, but the latter was published first.

[55] Listed with the Piltdown skull as one of two syntypes. It is here created the lectotype of the taxon.

[56] See my 1964 paper on this subject (op. cit.).

[57] See my summary of current opinion on Hominid classification in my article Quantitative Taxonomy and Human Evolution in *Classification and Human Evolution*, edited by S. L. Washburn, Chicago, 1963, p. 69.

[58] The relevant article in CINZ (23(e)) states that taxa formed by the union of two or more taxa take the oldest valid name among those of its components.

[59] It has been suggested that the name *Pithecanthropus* should be perpetuated as a sub-generic name. This usage, however, is of no taxonomic significance unless there are *more than two* species in the genus. The same comment applies to the use of *Zinjanthropus* as a sub-genus.

[60] The relationships and 'grades' shown in Table 2 have been modified from those published recently by Professors P. V. Tobias and G. H. R. von Koenigswald in a paper entitled 'A Comparison between the Olduvai Hominines and those of Java and some Implications for Hominid Phylogeny,' *Nature*, **204**, 515–18.

[61] Quoted from *The Discourses and Sayings of Confucius*, Ku Hung-Ming, Shanghai, 1898.

Notes

1. On the interpretation of the word 'indication' in Article 25(a).

   Generic names: Before 1931, the placing of a new generic name in immediate juxtaposition with an accepted specific name is considered to render that generic name valid as the genus of which the specific name designates the type species; that species is thus transferred to the new genus. Articles 11, 12, 16(a) (v). After 1930 the generic name is only valid under these conditions if the genus is monotypic unless the type species is clearly indicated. (E.g. *Plesianthropus* 1938 is valid as a new genus of which the only species is *Plesianthropus transvaalensis* (Broom 1936). See Article 13.)

   Specific names: New specific names are available if accompanied by a definite and unqualified citation of an earlier and valid name of the species for which the new name is proposed. Articles 11 & 16 (iii). The new name has not got priority at the time of its creation, but is available as an objective junior synonym. If a new name is published merely as a synonym it is not thereby made available (Article 11(d)).

2. Indication of the holotype of the taxon by means of a vernacular name is not a valid indication according to Article 16(b)(i).

3. *Small variations in spelling.*

   Generic names: Article 33 states that any unjustified emendation of a name has status in nomenclature and is a junior objective synonym of the name in its original form. But there are exceptions to this when an emendation is justified. Firstly, an incorrect subsequent spelling (I.S.S.) of an earlier generic name does not constitute a new genus

(Article 33(b)) (*Palaeanthropus* 1909 and *Palaeoanthropus* 1921 are considered to be a case in point). Secondly, a valid emendation is acceptable if it corrects an invalid original spelling (Article 33(a)(i)). (See Mugharet es-Skhul.)

Specific names: Article 58(8) states that the use of different connecting vowels in compound words of two specific names does not make them synonyms but homonyms. The endings -ensis and -iensis in geographical names are also homonymous (Article 58(ii)). (*Krapinensis* 1902 and *krapiniensis* 1911 are considered to be a case in point.)

4. Specific names do not bear capital letters even if they are originally given them (Article 28).

5. The Latin names of varieties are available as sub-specific names if introduced before 1961 (i.e. *Homo neanderthalensis* var. *krapinensis* Gorjanovic-Kramberger, 1902) (Article 17(9)).

6. Alternative names are often given together with valid names as being in some way more appropriate. Sometimes they are cited without qualification, but whether a reason is given or not they are usually presented for one of two reasons:

    (1) the new name is considered to be etymologically more appropriate, i.e. *H. primigenius* instead of *H. neandertalensis*. Names indicative of a particular geographical location become inappropriate when the range of the species is revealed.

    (2) The new name is chosen for reasons associated with a particular phylogenetic theory and with disregard of the rules, e.g. *Homo pithecanthropus* Manouvrier instead of *Pithecanthropus erectus* Dubois.

    Such alternative names are available as objective junior synonyms and are listed as such. They occur very commonly in the Hominidae. See also Note 1.

19

The paper by Ian Tattersall reprinted here represents part of a current trend among some paleoanthropologists toward recognizing more taxa in the hominid fossil record. This paper is concerned more with the true number of fossil species we should "expect" to recover and recognize than with more objective taxonomic matters such as what names should be applied to such fossil taxa. We expect the controversy implicit in Tattersall's paper to continue into the indefinite future of fossil hominid studies. What is a fossil species? How is it recognized and delimited? What macroevolutionary pattern(s) does the hominid fossil record reveal?

Tattersall is prepared to recognize more hominid species than are usually presented in introductory textbooks, whose taxonomies tend to be relatively "lumped," with few distinct taxa. Most textbooks of the last decade tend to recognize the same four species of *Australopithecus* (*A. afarensis, A. africanus, A. robustus, A. boisei*) and three species of *Homo* (*H. habilis, H. erectus, H. sapiens*). A point which is implicit in this paper was also clearly made by Campbell (selection 18): previously created taxon names, even those little-used and usually ignored, retain potential usefulness in new or revised classifications which recognize either more species, or more subspecies, than current ones.

Species Recognition in Human Paleontology

Ian Tattersall

Over the past several years increasing attention has been paid to the search for patterns in the human fossil record (e.g., Cronin *et al.*, 1981; Rightmire, 1981; Eldredge & Tattersall, 1982). The reliability of any attempt to recognize pattern, however, is constrained by the accuracy with which we are able to recognize species in that record; this is true whether one subscribes to the view that species are objective entities in time as well as in space, or, more traditionally, holds that the practical problem is to distinguish between separately evolving lineages within which species are ephemeral in time, if not in space. Irrespective of one's position on the nature of the evolutionary process, species or at any rate time-successive lineages of species, are the units with which one has to deal in trying to unravel the sequence of evolutionary events. The delineation of these units is thus of critical concern. My aim in this short essay is not to contribute to the debate over evolutionary mechanism, which will doubtless well outlive this century without my help. Rather it is, first, to focus attention upon the observation that whatever the exact nature of currently popular criteria of species recognition in paleoanthropology (and they are rarely if ever articulated), to judge by their results they are unrealistic; and, second, to point out some of the consequences of this for our interpretations of the later stages of human evolution.

In their seminal paper on Miocene hominoids, Simons & Pilbeam (1965) came about as close as any paleoanthropologist has yet come to precision on the question of how one proceeds to sort assemblages of fossils into discrete groups: "In order to establish a valid species it should be necessary to show characters in the available fossil

From Ian Tattersall, "Species Recognition in Human Paleontology," *Journal of Human Evolution*, Vol. 15, pp. 165–175 (1986). Reprinted by permission of Harcourt, Brace, and Co., London, and the author.

material which purport to be of the same magnitude as those which separate related living species" (p. 101). But although few have seen fit to quarrel with it this sage advice has been honored almost exclusively in the breach, even by its authors themselves. For while Simons & Pilbeam specifically urged the evaluation of inter-species variation, recent paleoanthropological practice has been to focus attention on intra-species variation. The question is, in effect, most commonly rendered: "do the limits of variation we see in our fossils exceed the limits of variation we see in samples drawn from extant primate species?" Indeed, the years since Simons & Pilbeam wrote have seen a steady flow of theses that document the ranges of variation shown by numerous characters, mostly dental, in ape and human skeletal samples. Almost invariably, such studies have been aimed at establishing norms against which the variation seen in, say, australopith samples may be evaluated. Hand in hand with this concern for variation has gone the triumph of the lumping ethic which, in laudably rejecting the typological thinking that bedeviled paleontology in the earlier years of this century, has in certain cases proceeded with an equal lack of realism to the opposite extreme.

It is not hard to see why intraspecies variation has proven so much more beguiling to paleoanthropologists than has interspecific variation. For, doctrine of common sense though Simons & Pilbeam's exhortation appears to be, it offers no practical guide to the sorting-out of fossil assemblages. The reason for this is, of course, that there is no direct relationship, indeed no consistent relationship at all, between speciation and morphological change (e.g. Vrba, 1980). On the one hand, speciation may take place in the absence of appreciable morphological divergence, while on the other considerable local differentiation may occur, particularly within geographically widespread species, without any rupture of the reproductive continuity that both binds and limits the species population. Thus, while few would disagree that there has been no more significant theoretical advance in systematics than the shift from the morphological to the biological concept of species, this change has opened the door to a host of practical problems when it comes to the actual interpretation of the fossil record. For quite apart from the considerable difficulty of translating magnitudes of variation from one species or species pair to another, it is also evident that such taxonomic decisions are critically affected by the choice of which living species—or pair of related species—is to be taken as arbiter of the variability permissible in a fossil assemblage recognized as a species. There is no guarantee that the "closest living relative" (as it usually boils down to, whether or not this is determinable with any certainty) will be the most appropriate model to apply; for instance, a species occupying a limited relict distribution (one of the great apes, perhaps?) would almost certainly be an unreliable guide to the delineation of a widely-distributed related fossil species. Moreover, and much more importantly, almost invariably what is evaluated is *within-species* variation, which will be found to a greater or a lesser extent in all anatomical systems. But what is important in distinguishing among species is *between-species* variation. Where species are closely related (and hence most likely to be confused in the fossil record), variation of this kind is likely to be restricted to a few characters only; the vast majority of anatomical characters will overlap substantially or totally.

Despite the lack of association between taxic and morphologic evolution, however, a certain amount of generalization from the living fauna is possible as a broad guide to systematic practice in analyzing the fossil record. For speciation rarely, if ever, involves quantum morphological leaps. If one surveys the primates as a whole, one finds that the morphological differences between closely related species (say species usually classified within the same genus or subgenus) are commonly small, and restricted to only one or a few characters. Especially where such species are very closely related, for instance where two species are descended from the same parental species, distinguishing features are often limited to soft tissue characteristics which do not preserve in the fossil record. And even in those cases where distinctions in certain hard-tissue characters are observable between closely related species, substantial or complete overlap would be expected in all others. Because of this it is easy for the paleontologist to confuse variations of the two kinds, especially when the time arrives for taxonomic judgement on small samples of fossils. The keener one's appreciation of the important fact of intra-species variability, firmly and justifiably fixed as the keystone of modern paleoanthropological systematics, the greater the danger of doing this becomes. Clearly, every species is variable, and this central reality cannot be ignored in comparing two fossils; for just as similarity in a given set of features does not guarantee that two forms belong to the same species, observed differences may not imply species distinction. Nevertheless, each species varies within limits, and it is hard to avoid the conclusion that under current taxonomic practice there is a distinct tendency to underestimate the abundance of species in the primate, and notably the hominid, fossil record.

This observation runs counter to Templeton's (1982) speculation that since mammals are more prone than other vertebrates to form polytypic species, the number of mammal species in the fossil record is likely to be overestimated. Templeton's argument, however, is based purely on presumed relative propensities of different population structures to promote rapid population divergence, hence polytypism, which he apparently equates with detectable morphological shifts. But as we have seen, it is unrealistic to make this equation, and morphology itself, certainly those aspects of it that preserve in the fossil record, tells a different story. Cranioskeletal differences between primate subspecies of the same species tend to be tiny, if observable at all. For example, differences on the order of those separating the various subspecies of *Propithecus diadema* would certainly be missed by a paleontologist with only teeth and bones to examine, especially if he had a keen sense of intra-population variation. Even the highly polytypic extant species *Homo sapiens* varies much less than one would infer from the variety of fossil morphs that has been crammed into it.

The tendency to underestimate low-level taxic diversity in the fossil record is, it must be said, infinitely preferable to the opposite fault, widespread in earlier days, of baptizing each new fossil specimen with its own name. But it may nonetheless subtly give rise to a destructive misconception. It may quite persuasively be argued that it is unimportant whether the species we recognize in the fossil record are "real" species or not. As long as those units we recognize as species are "natural," or monophyletic, does it really matter whether they consist of individual species or of aggregates of closely related species? Where there is no reasonable prospect of distinguishing such putative species on the basis of preserved materials there can, of course, be only one reasonable answer to this question. Nonetheless, where respect for intra-species variability leads to the inclusion within a single species of fossils representing more than one readily identifiable morph, it is absolutely crucial that we be totally confident that only a single species is in fact involved. This is no small matter. For while species are objective historical entities that we are obliged to confront, morphological varieties *within* species are acknowledged to be no more than epiphenomena, whose individual historical identity has not been established.

Species, reproductively isolated from the rest of the living world, acquire their discreteness as the result of an irreversible genetic and historical process. Subspecies, on the other hand, while distinguishable in some way and at some point, are in concept ephemera, unbounded by reproductive barriers; their unique identity may at least potentially be lost at any time in merging with other subspecies populations. Only with speciation, with the development of reproductive isolation, does the identity of what began as a subspecies become definitively established. Of course this does not necessarily mean that subspecies, as entities, do not demand study and explanation; they have origins and histories just as species do. But the fact remains that, to paleontologists in general, subspecies are epiphenomena which do not merit the attention paid to species: it is species, not subspecies, that are the units of evolution, or at least of evolutionary study. And indeed, to judge from the generally insignificant intertaxic diversity in hard parts exhibited at the lowest taxonomic levels in the modern primate fauna, the pursuit of subspecies in the fossil record is at best fraught with difficulty, and is more probably futile. This is why it is critical to avoid relegating distinct morphological variants observed in the fossil record to the status of subspecies—or their informal equivalent—without the best of reasons for doing so; for, as mere subspecies, even clearly identifiable morphs do not urgently demand analysis and explanation. Indeed, it might well be argued that it would be better for the comprehensiveness of our understanding of the human fossil record that, if err we must, we err (within reason!) on the side of recognizing too many rather than, as is the tendency, too few species units. After all, to hark back to an earlier comment, even a subspecies has a history worthy of investigation.

All this, of course, begs the problem of distinguishing separate morphs where differences are slight, while the nonconcordance of taxic with morphological change seems to eliminate any objective solution to the practical problem of species recognition. What I wish to emphasize here, however, is that where distinct morphs can readily be identified it would seem most productive to assume that they represent species unless there is compelling evidence to believe otherwise. To brush morphological diversity under the rug of an all-encompassing species is simply to blind oneself to the complex realities of phylogeny.

The Human Fossil Record

By 1950, at a time when the known hominid fossil record was vastly more limited than it is at present, the names of over a dozen genera, and many more species, were commonly encountered in the less than voluminous literature on fossil Hominidae. In that year, however, a long overdue and highly influential paper by Ernst Mayr started the lumping bandwagon rolling with the declaration that: "the acceptance of the new concept of biologically defined polytypic species necessitates the upward revision of all other categories" (Mayr, 1950, p. 110). Mayr spurned the idea of monotypic genera as leading to an "inequality of categories," and at that time he even argued for including *Australopithecus* in *Homo*, since "within this type there has been speciation resulting in *Homo sapiens*." Nonetheless, with infinitely clearer vision than those of his successors who raised the "single species hypothesis" to an article of faith, Mayr explicitly recognized not only that to arrive at this taxonomic conclusion required "a deliberate minimizing of the brain as a decisive taxonomic character," but also that the necessary assumption that "all of these forms, including *Australopithecus*, are essentially members of a single line of descent," was exceedingly fragile, for "Additional finds might readily disprove this." (pp. 110–111.) Subsequent paleoanthropological discoveries have emphasized the wisdom of Mayr's caveat, but the three and a half decades following his salutary contribution have witnessed the ascendance of what has finally become a destructively minimalist approach to hominid systematics.

Today, happily, the realization has begun to dawn that a significant degree of taxonomic variety exists in the remoter stages of hominoid history, among the australopiths. The genus *Homo*, however, which following Mayr's exhortation has been expanded to embrace over two million years of time and a large array of distinct hominid morphs, is commonly viewed as containing only three species: *Homo habilis, H. erectus*, and *H. sapiens*. The first of these poses acute problems of definition and recognition, exacerbated by the fact that the only specimens in a motley collection that are truly distinctive morphologically are not associated with the name. The second has, in the years since Mayr wrote, expanded beyond its vague role as the "man of the middle Pleistocene" to include a diverse array of fossils that future scholars will surely find ample cause to subdivide—unless then can contrive to maintain the fiction of the "grade" to keep this diversity decently obscured. The interpretation of most human fossils subsequent to about 0.5 myr as belonging somehow to *Homo sapiens* is perhaps the most comprehensive smokescreen of all, and could only have been maintained by a zealous refusal to consider characters other than brain size. I have suggested elsewhere (Eldredge & Tattersall, 1982) that the urge to include forms as diverse as Petralona, Steinheim, Neanderthal and you and me in the single species *Homo sapiens* must be sociological in origin—there is something vaguely disgraceful about discriminating against other large-brained hominids—and I do not wish to beat that horse further here; but the fact remains that the only rational explanation for the taxonomic corralling together of these widely differing fossils (for age will not do it, Wolpoff, 1980, to the contrary) is the setting of an unconscious "cerebral Rubicon," perhaps at somewhere around 1200 ml. The fact that such a Rubicon, unconscious or otherwise, would exclude a good many members of the living species from *Homo sapiens* has gone unremarked. And while, if my surmise is correct, one can only applaud the generous liberal sentiment that leads to the inclusion in *Homo sapiens* of all hominids whose brain size falls comfortably within the modern range, one is equally obliged to point out that to conclude that no species differences exist within this array of hominids is to confuse the two kinds of morphological variation. The huge variation in normal brain size found among modern humans exemplifies within-species variation, and is largely or entirely irrelevant to the issue of taxic diversity. I have already suggested that closely-related primate species will show identical or substantially overlapping ranges of variation in almost all morphological features; and while it could, I suppose, be argued that both the lower-than-modern average brain size of "archaic *Homo sapiens*" and the higher-than-modern average brain size of the classic Neanderthals are features that could be used in demarcating both from *Homo sapiens*, what is much more important in weighing the question of how many hominid species we are faced with in the middle and late Pleistocene is the constellation of features in which both clearly differ from each other and from ourselves.

Even a relatively cursory survey of the spectrum of living primates is sufficient to reveal that modern *Homo sapiens*, by now numbering some four billion individuals and for long represented in virtually every inhabitable region of the earth, is a rather highly polytypic species. Nonetheless, like every other species, polytypic or otherwise, *Homo sapiens* is a cohesive biological entity which, while variable, is far from infinitely so. Regressing in time, we find that back to about 30–35 kyr ago more or less all human fossils fall into the morphological range bracketed by the living species. Some subspecies of *Homo sapiens* may well, of course, have emerged within that period only to become extinct, and thus it would be reasonable to expect to find, as indeed we do, the occasional *Homo sapiens* fossil that slightly exceeds this range in one character or another. If we look further back, however, we abruptly lose the ability to fit the human fossils we find into this frame of variation, with the exception of a few pieces that are either fragmentary or uncertainly dated. But although almost all human fossils prior to this time are very different in aspect from anything we see in our very variable living species, they are nonetheless regularly classified as *Homo sapiens*. With the proviso, to be sure, that they are "archaic," or otherwise different; but *Homo sapiens* nonetheless.

Rarely indeed, however, have paleontologists ever found it necessary to distinguish between "archaic" and "anatomically modern" types of the same species, and there seems scant justification for squeezing these distinct hominid morphs into a single species. In any group other than Hominidae the presence of several clearly recognizable morphs in the record of the middle to upper Pleistocene would suggest (indeed, demonstrate) the involvement of several species. Any mammalian paleontologist seeing morphological differences on the order of those separating modern humans from their various precursors, and the latter from each other, would have no difficulty in recognizing a number of separate species. And in this decision there is no obvious place for special pleading, even where it is our own closest relatives that are involved. It is not in the least surprising that various authors, for example Stringer *et al.* (1979), Wu (1981) and Rightmire (1983), have been able to discover derived features shared with their successors in such fossils as Kabwe, Ndutu, Petralona, Arago and Dali. This, however, is no argument for conspecificity when they also show differences of a magnitude that in any other primate family would be accepted without demur as demarcating separate species. Sisters there may be among these morphs; conspecifics, no.

It is not my intention here to go into detail on the morphological justifications for recognizing several separate species in the post-*Homo erectus* human fossil record. Indeed, even if this contribution were not cast as an essay rather than as a research report, this would be impractical for two interrelated reasons. In the first place, despite the existence of an extensive literature, discussion of the detailed morphological evidence for a multiplicity of human species over the past half-million years is difficult without extensive study of the actual fossil specimens involved. This is because studies undertaken to date have tended to focus on the continuities between, say, "archaic *H. sapiens*" and temporally contiguous species—*Homo erectus*, the Neanderthals, or "anatomically modern *Homo sapiens*"—and hence, presumably, largely on primitive characters. Most of the literature, indeed, though crammed with comparisons, is extraordinarily uninformative if one wishes to know if the character states being compared are primitive or derived, or how widely they are shared. Secondly, except very recently in the case of the Neanderthals, few attempts have been made to define distinctive later hominid groups as morphs, let alone as species, not only as a result of the search for continuity but also, one presumes, in consequence of their epiphenomenological status as intraspecific variants. One should note, however, that with the adoption by some paleoanthropologists of a more or less cladistic approach to the reconstruction of phylogeny, there have recently been some welcome positive moves. Thus Hublin (1978) listed a set of apomorphies of the skull rear among Neanderthals, and Day & Stringer (1982) proposed "working definitions" of the living species *Homo sapiens* and the Afro-Asian species *Homo erectus*. Most recently still, Stringer, *et al.* (1984) have provided lists of "anatomical characters of modern *Homo sapiens*" (Table 1, p. 54) and "proposed Neandertal autapomorphies and common characteristics" (Table 2, p. 55), and Vandermeersch (1985) has also couched his discussion in these terms. Encouragingly, the contemplation of lists of this kind has led Stringer (1984a, 1985) to speculate on whether the name *Homo sapiens* might not best be reserved for humans of modern aspect, "extreme" as he felt the suggestion would appear. In any event, for the purposes of the present discussion it is sufficient to point out that several clearly different morphs of "*Homo sapiens*," each already considered sufficiently distinct to require an adjectival qualifier, are widely recognized in the fossil record, in addition to a considerable number of individual specimens whose status it has not generally been felt necessary to examine in detail given the umbrella existence of *Homo sapiens*.

To begin with, of course, there is "anatomically modern" *Homo sapiens*, with an impressive set of autapomorphies as listed by Day & Stringer (1982): cranium short but high; parietal arch long and high, narrow inferiorly and broad superiorly; frontal bone high; supraorbital torus not continuous, but divided into lateral and medial portions; occipital bone well curved, not angulated; mental eminence on mandible; gracile limb bones. Even as a preliminary approximation these characters reflect a morphology clearly distinct at the species level from that of any other hominid form. Although archaeological evidence does not strictly provide taxonomic characters, one might add that recognizing the distinctness of *Homo sapiens* in this restricted sense makes it much easier to understand the extraordinary behavioral advancements (see, for example, White, 1982, and Klein, 1985) that apparently accompanied the appearance on earth of people of our own kind following a long period that witnessed relatively little technological or behavioral innovation.

As distinct as *Homo sapiens* is *Homo neanderthalensis*. "Definitions" of this group have traditionally amounted to descriptions, intermingling primitive and derived characters, but clarifications have recently begun to appear. Thus among the Neanderthal apomorphies identified by Hublin (1978), Stringer *et al.* (1984), and Vandermeersch (1985) are the following: lambdoid flattening and "bunning" of the occiput; the mid-sagittal division of the occipital torus, with formation of a suprainiac fossa; the presence of an anterior mastoid tubercle and the conformation of the surrounding region as detailed by Vandermeersch (1985); the rounded ("en bombe") form of the cranium in posterior view; pronounced midfacial prognathism and its correlates, including a retreating zygomatic profile and retromolar spaces; nasal aperture capacious, with nasal bones projecting; relatively wide sphenoidal angle; extreme breadth of the anterior teeth. To listed postcranial characteristics such as the long and plate-like superior pubic ramus might be added pronounced radial curvature, flattening of the first carpometacarpal saddle joint, and phalangeal length proportions.

The longstanding dispute over whether the group of hominid fossils containing the Arago material should be classed as *Homo sapiens* or *Homo erectus* reflects the underlying reality that this form, most commonly referred to by Anglophone paleoanthropologists as "archaic *Homo sapiens*," actually belongs with neither species. Specimens such as those from Bodo, Kabwe, Petralona and Arago (*Homo heidelbergensis*, perhaps?) compose a variable morph (recognized in its having been named, however inappropriately) that is clearly demarcated at the species level from *Homo erectus*, from *Homo neanderthalensis*, and from *Homo sapiens*. Stringer (1985) has summarized several of its distinguishing characters in a Venn diagram which emphasizes the need for further analysis of the apomorphies of the group. One might point out, however, that except in the form of its supraorbital torus, this hominid species appears distinct also from that best represented by the Steinheim and possibly Swanscombe specimens (as Stringer, 1985, has also implied in commenting on characters of the rear of the cranium). Both Hublin (1982) and Vandermeersch (1985) have pointed to the presence in the Steinheim skull of a character interpretable as a suprainiac fossa, and have suggested that Steinheim represents a proto-Neanderthal; Stringer (1985) has, however, rejected this interpretation, while suggesting nonetheless that this form departs from the Neanderthals largely in pleisomorphous character states. Interestingly, most speculations on Steinheim's links with the Neanderthals has focused on the rear of the skull, while those who have discovered "Neanderthal-like" features in the Petralona-Arago group have pointed most convincingly to characters of the face (see Vandermeersch, 1985).

This highly incomplete listing of characters, specimens and morphs is intended merely to be illustrative, and clearly a great deal of work remains to be done in sorting out the number of distinct morphs represented in the geographically scattered hominid fossil record of the Middle and Late Pleistocene. However, it does demonstrate that at least three or four (and quite probably more if the entire spectrum of material were considered) separate hominid groups can be distinguished among the "*Homo sapiens*" fossils of this time. Moreover, it strongly suggests that, when measured by the standards that apply to other primates the morphological differences between these groups amply justify species distinctions. If my operational criteria for species recognition appear somewhat informal this is necessarily the case, in consequence of the lack of correlation between speciation and degree of morphological change; but as I have already argued, morphological differences of the degree discussed invariably indicate among living forms that speciation has occurred.

Rightmire (1985) has similarly made the point that, if species are to be viewed as real in time as well as in space, it is necessary to recognize four species, *Homo erectus*, *Homo heidelbergensis*, *Homo neanderthalensis* and *Homo sapiens*, in the middle to late Pleistocene human fossil record. He found this scheme wanting, however, in that many or all later Pleistocene hominids from Africa and China do not clearly fit into one or another of these species, and preferred Bonde's (1981) grouping of all as subspecies of *Homo sapiens*. But so to conclude is once again to use the device of an all-embracing *Homo sapiens* to obscure variability that actually exists, and requires analysis. If African and Chinese hominids of this period show apomorphies separating them from the essentially European scheme on which this division into species is based, it is certainly appropriate at least to enquire whether yet more species might be involved. Certainly the point is already conceded that more morphs are present than can be embraced by the European scheme, and these morphs require analysis and explanation of the kind they are most likely to be denied if they are dismissed as representatives of "*Homo sapiens*" in some undefined larger sense.

Recognizing a multiplicity of hominid species in the mid-to-late Pleistocene will, of course, go against the grain for many paleo-anthropologists. After all, there is a beauty in linear simplicity that the aesthetes among us, in particular, would be reluctant to see vanish. Nonetheless, this systematic conclusion is hardly surprising when one considers the enormous climatic and sea-level fluctuations that marked the period. Without any doubt, and most especially in temperate latitudes, these fluctuations fragmented and recoalesced hominid populations, thereby providing ideal conditions for speciation and competition. The potential that existed during the middle and late Pleistocene for the local divergence of populations is even recognized in models such as that adopted by Wolpoff *et al.* (1984), who go on to posit longstanding continuities within regional human populations. However, if one accepts that different environments in different regions can produce morphological discontinuities between populations, then one must also accept that environmental change *in situ* can do the same. Perceived long-term morphological trends in local human populations over this period then become more difficult to defend, even when lavishly tricked out with genetic sophistry. Whatever reasonable model we accept of the pattern of evolution in the later part of the hominid fossil record, it is clear that not all of the hominid variants of the mid-to-late Pleistocene can have been ancestral to our own species, and that the origin of *Homo sapiens* is to be sought in a morph whose own origin was linked to a specific geographic region. Given the anatomical range already known among such morphs it is virtually certain that this population represented a distinct hominid species which coexisted, at least temporally, with others.

The discussion of the greater spectrum of middle to late Pleistocene human fossils, from sites worldwide, inevitably spills over into the problem of defining *Homo erectus*, recently ably discussed by Andrews (1984), Rightmire (e.g. 1984, 1985), Stringer (1984b), and Wood (1984). Stringer, in particular, makes the point that African and Asian *Homo erectus* can be assigned to a "grade" called *Homo erectus*, but not to a species of that name. However, like it or not, *Homo erectus* is a species name; "grades" are not recognized in the International Code of Zoological Nomenclature, and neither, in any objective sense, are they found in nature, Wolpoff *et al.* (1984) to the contrary notwithstanding. The "grade," indeed, is one of the most destructive canards that paleoanthropology has ever seen fit to inflict upon itself: a meaningless and undefined concept, apparently leaning heavily on brain size, that can be used to entomb all kinds of morphological loose ends and thus eliminate the need to examine them. As long as we apply standard systematic terminology to our fossil record, we should be clear about the meaning of the terms we use, and we should be wary of introducing others that serve only to obfuscate. The authors named above have elegantly shown that characters traditionally used to define *Homo erectus* include a mishmash of primitive and derived character states, and that the African and Asian forms are not distinguished by the same apomorphies. Moreover, Andrews makes the important point that comparisons of *Homo erectus* with other hominid species have been largely with *Homo sapiens* alone. Only when species have been adequately defined morphologically can appropriate comparisons be made, and the distribution of character states across species can be used to generate phylogenetic hypotheses. Lumping diverse morphologies into "grades" kills this process before it can start.

In this note I have attempted to raise questions rather than to solve them, and above all to draw attention to the fact that by masking the significance of much observed morphological variation, inadequate concepts of species delineation have in the past impeded advancements in our understanding of hominid phylogeny. Although it appears difficult or even impossible to arrive at an objective, or universally applicable, solution to the problem of species recognition in the fossil record, cognizance of the general patterns of between-species morphological variation in the living biota will help us better to assess that record. That we do so is crucial, for we will make little useful progress in human evolutionary studies until we adopt more realistic working criteria of species recognition, and define those species we recognize in terms of derived, or at least of distinct, morphologies. As I have noted, this is so even under a gradualist model of evolution; for however one views the evolutionary process, paleoanthropology is a comparative science in which rational recognition is necessary of the units to be compared. Only when we have realistically defined the species we accept we will be able to broach the problem of the relationships between such species, whether linear or collateral, and then in turn proceed towards an understanding of larger-scale patterns in human evolution.

Summary

The search for patterns in the human fossil record has been handicapped by the inadequacy of the criteria used to recognize species in that record. The fact that taxic and morphological change need not be associated directly means that there can be no objective means of sorting fossils into species. In general, however, closely related primate species show only minor morphological differences from one another, sometimes in parts of the anatomy that do not preserve in the fossil record. There is thus a tendency to underestimate species diversity in the fossil record; and while lumping may appear sophisticated because it is avowedly anti-typological, too much of a good thing has in the end become a liability to our interpretation of the substantial morphological diversity that exists in the human fossil record. Subspecific categories are regarded as epiphenomena that do not justify the attention accorded to species, so it is critical that where distinct morphs exist in the fossil record they are not relegated to subspecific status without good reason. Several morphs are distinguishable in the later part of the human fossil record that are distinct both from each other and from the living species *Homo sapiens*. These include *Homo neanderthalensis*, *Homo heidelbergensis*, and probably also *Homo steinheimensis*. The recognition of morphs in the later part of the human fossil record has been based largely on the assemblage of fossils from Europe, and many fossils from Asia and Africa do not fit the European scheme. Probably other species will prove identifiable from other areas of the world when the appropriate comparisons are made. The existence of a multiplicity of hominid species in the middle to late Pleistocene around the world would not be surprising in view of the great climatic and eustatic oscillations, doubtless involving the frequent fragmentation and recoalescence of human populations, that marked the period.

Acknowledgements

This article was improved by the comments of Eric Delson, Niles Eldredge, and three anonymous reviewers, none of whom necessarily agrees with anything it contains. I am indebted to the Richard Lounsbery Foundation for its generous support. This is Contribution no. 18 of the Lounsbery Laboratory of Biological Anthropology, American Museum of Natural History.

References

Andrews, P. (1984). An alternative interpretation of characters used to define *Homo erectus*. *Cour. Forsch. Inst. Senckenberg* **69**, 167–175.

Bonde, N. (1981). Problems of species concepts in paleontology. In (J. Martinelli, Ed.) *Concept and Method in Palaeontology*, pp. 19–34. Barcelona: University of Barcelona.

Cronin, J. E., Boaz, N. T., Stringer, C. B. & Rak, Y. (1981). Tempo and mode in hominid evolution. *Nature* **292**, 113–122.

Day, M. H. & Stringer, C. B. (1982). A reconsideration of the Omo Kibish remains the *erectus-sapiens* transition. *Premier Congresse Internationale de Palaeontologie Humaine* Prétirage, Vol. 2, pp. 814–846.

Eldredge, N. & Tattersall, I. (1982). *The Myths of Human Evolution*. New York, Columbia University Press.

Hublin, J. J. (1978). Quelques caractères apomorphes du crane néandertalien et leur interprétation phylogénique. *C. R. Acad. Sci. Paris.* D, **287**, 923–926.

Hublin, J. J. (1982). Les anténéandertaliens: présapiens ou prènèandtertaliens? *Geobios*, Mém. Spéc. **6**, 345–357.

Klein, R. G. (1985). Breaking away. *Nat. Hist.* **94**(1), 4–7.

Mayr, E. (1950). Taxonomic categories in fossil hominids. *Cold Spring Harb. Symp. Quant. Biol.* **15**, 109–118.

Rightmire, G. P. (1981). Patterns in the evolution of *Homo erectus*. *Paleobiology* **8**, 241–246.

Rightmire, G. P. (1983). The Lake Ndutu cranium and early *Homo sapiens* in Africa. *Am. J. Phys. Anthrop.*, **61**, 245–254.

Rightmire, G. P. (1984). Comparisons of *Homo erectus* from Africa and southeast Asia. *Cour. Forsch. Inst. Senckenberg* **69**, 83–98.

Rightmire, G. P. (1985). The tempo of change in the evolution of mid-Pleistocene *Homo*. In (E. Delson, Ed.) *Ancestors: The Hard Evidence*, pp. 255–264. New York: Alan R. Liss.

Simons, E. L. & Pilbeam, D. R. (1965). Preliminary revision of the Dryopithecinae (Pongidae, Anthropoidea). *Fol. Primat.* **3**, 81–152.

Stringer, C. (1984a). Human evolution and biological adaptation in the Pleistocene. In (R. Foley, Ed.) *Hominid Evolution and Community Ecology*, pp. 55–83. London: Academic Press.

Stringer, C. B. (1984b). The definition of *Homo erectus* and the existence of the species in Africa and Europe. *Cour. Forsch. Inst. Senckenberg* **69**, 131–143.

Stringer, C. B. (1985). Middle Pleistocene hominid variability and the origin of late Pleistocene humans. In (E. Delson, Ed.) *Ancestors: The Hard Evidence*, pp. 289–295. New York: Alan R. Liss.

Stringer, C. B., Howell, F. C., & Melentis, J. K. (1979). The significance of the fossil hominid skull from Petralona, Greece. *J. Arch. Sci.* **6**, 235–253.

Stringer, C. B., Hublin, J. J., & Vandermeersch, B. (1984). The origin of modern humans in western Europe. In (F. H. Smith & F. Spencer, Eds.) *The Origins of Modern Humans: A World Survey of the Fossil Evidence*, pp. 51–135. New York: Alan R. Liss.

Templeton, A. R. (1982). Genetic architectures of speciation. In (Barigozzi, C., Ed.) *Mechanisms of Speciation*, p. 105–121. New York: Alan R. Liss.

Vandermeersch, B. (1985). The origin of the Neandertals. In (E. Delson, Ed.) *Ancestors: The Hard Evidence*, pp. 306–309. New York: Alan R. Liss.

Vrba, E. S. (1980). Evolution, species and fossils: How does life evolve? *S. Afr. J. Sci.* **76**, 61–84.

White, R. (1982). Rethinking the Middle/Upper Paleolithic transition. *Curr. Anthrop.* **23**, 169–192.

Wood, B. (1984) The origin of *Homo erectus. Cour. Forsch. Inst. Senckenberg* **69**, 99–111.

Wolpoff, M. H. (1980). *Paleoanthropology*. New York: Knopf.

Wolpoff, M. H., Wu, X., & Thorne, A. G. (1984). Modern *Homo sapiens* origins: a general theory of hominid evolution involving the fossil evidence from East Asia. In (F. H. Smith & F. Spencer, Eds.) *The Origin of Modern Humans: A World Survey of the Fossil Evidence*, pp. 411–483. New York: Alan R. Liss.

Wu, X. (1981). A well-preserved cranium of an archaic type of early *Homo sapiens* from Dali, China. *Sci. Sin.* **24**, 530–541.